AF594834

UNDESIRABLE

UNDESIRABLE

Invasive Species, Humans, and Where We All Belong

AMY CRAWFORD

OXFORD
UNIVERSITY PRESS

Oxford University Press is a department of the University of Oxford.
It furthers the University's objective of excellence in research, scholarship,
and education by publishing worldwide. Oxford is a registered trade mark of
Oxford University Press in the UK and in certain other countries.

Published in the United States of America by Oxford University Press
198 Madison Avenue, New York, NY 10016, United States of America.

© Amy Crawford 2026

All rights reserved. No part of this publication may be reproduced, stored in a retrieval system, transmitted, used for text and data mining, or used for training artificial intelligence, in any form or by any means, without the prior permission in writing of Oxford University Press, or as expressly permitted by law, by license or under terms agreed with the appropriate reprographics rights organization. Inquiries concerning reproduction outside the scope of the above should be sent to the Rights Department, Oxford University Press, at the address above.

You must not circulate this work in any other form
and you must impose this same condition on any acquirer.

CIP data is on file at the Library of Congress.

ISBN 9780197768082

DOI: 10.1093/oso/9780197768082.001.0001

Printed by Sheridan Books, Inc., United States of America

The manufacturer's authorized representative in the EU for product safety is
Oxford University Press España S.A. of Parque Empresarial San Fernando de Henares,
Avenida de Castilla, 2 – 28830 Madrid (www.oup.es/en or product.safety@oup.com).
OUP España S.A. also acts as importer into Spain of products made by the manufacturer.

To my parents, who made it all possible.

CONTENTS

Introduction

MONSTER STORIES

Nailed to the side of the concession building at the public beach where I take my kids in Pinckney, Michigan, is what amounts to a wanted poster.

For swans.

"You can't see the ugliness of mute swans by looking at them," reads the notice, placed by the state's Department of Natural Resources, or DNR, which goes on to compare the regal birds' feet to toilet plungers, "swirling out substrate and degrading water quality." Worse, they "consume 8–10 pounds of vegetation a day, ripping plants out by their root. They pollute ponds and foul beaches with their copious feces." Little can be done about these "Merciless Killers," laments the DNR, noting that efforts to stop the swans' population growth through regulated hunting programs and by injecting bleach into their eggs have "been shouted down and blocked by the public. . . . Their breathtaking beauty has successfully disguised the true ugliness of this beast."

Mute swans' real crime, however, is that they originated in Europe, where, incidentally, their feeding, defecation, and mating habits have never been seen as a problem.

To be more precise, the ancestors of the birds that now paddle around Michigan's lakes and ponds were imported during the Gilded Age by Americans who sought a touch of Old World elegance for public parks and country estates. Finding their new homes to be quite comfortable, these supersized waterfowl soon reproduced and expanded their territory, until, in the early 20th century, they reached the upper Midwest. By that time, the region's native trumpeter swans, long hunted for their meat, skin, and feathers, were teetering on the brink of extinction.

Today, mute swans (*Cygnus olor*) thrive along the East Coast and around the Great Lakes, despite the efforts of Michigan's DNR and other conservation-focused agencies to, at minimum, destroy their reputation. But although they face new competition from their Eurasian cousins, trumpeter swans (*Cygnus buccinator*) have also made a stunning recovery.

In the 1950s, aerial surveyors discovered a remnant population of several thousand trumpeters in a remote part of southern Alaska. Thirty years later, biologists from Wisconsin, Minnesota, and Michigan flew there on small planes to gather hundreds of trumpeter swan eggs for use in a reintroduction program that would become one of the conservation movement's biggest success stories. The eggs were given to mute swan foster parents or hatched in incubators. In some cases human caregivers, who were fastidious about keeping these animals wild, wore swan costumes to ensure their charges wouldn't imprint on people. Captive breeding continues today, even as trumpeter swans have once again become fairly common—and notwithstanding some ecologists' concern that they may have spread beyond the bounds of their own historic range.

The mute swan and the trumpeter are similar in size, appearance, and lifestyle, and in their impact on the wetlands where they live. Like mute swans, trumpeter swans uproot large quantities of aquatic plants, and they defecate at least as much as their Eurasian cousins. From a distance, it can be hard for even an experienced birdwatcher to tell them apart (the difference lies in the color of the bill and the curve of the neck; the trumpeter is also slightly larger). Does it make sense that one bird should be reviled by conservationists while the other is painstakingly protected?

The poster at my local beach has, truth be told, always cracked me up a little. It certainly hasn't had the desired effect: Over the years, we have sometimes spotted a pair of mute swans breeding wantonly near the lake's far shore, apparently unaware that they're the targets of such strongly worded propaganda. The kids and I give them space—all swans can be aggressive when nesting—and the birds, in turn, stay well away from the bathing beach. "Merciless Killers" though they may be, the mute swans' reserve contrasts flatteringly with the behavior of native birds in the vicinity, including the ring-billed gulls (*Larus delawarensis*) that shriek and dive at picnickers and the Canada geese (*Branta canadensis*) that hiss at children.

But although the Michigan DNR's description of mute swans—"invading armies," "destructive force"—may sound absurd, even comical, when applied to a bird that has historically symbolized romance, authorities routinely deploy this sort of rhetoric against all kinds of non-native species.

Around here, mute swans are only one of many unwanted organisms that my family is warned of when we spend any time outdoors. At boat launches, the DNR urges us to "help stop aquatic hitchhikers" like hydrilla (*Hydrilla verticillata*) and zebra mussels (*Dreissena polymorpha*) by wiping down our kayaks and properly disposing of live bait. At trailheads, we're asked to "give invasive species the brush-off" by cleaning the treads of our boots, lest we spread the seeds of garlic mustard (*Alliaria petiolata*) or Japanese barberry (*Berberis thunbergii*). Everywhere, it seems, our forests, lakes, and streams are being invaded by foreign pests—and only we have the power to stop them.

In the war on invasive species, however, Michigan is far from the only front.

A Global Conflict

"Non-native," "noxious," "alien"—whatever we call them, plants, animals, and pathogens from elsewhere are a major concern of governments, farmers, and conservation groups around the world. And not without reason. For one thing, they cost humans a lot of money. According to a running estimate maintained by French biologists, between direct damage, reduced agricultural yields, and the expenses of mitigation, over the past fifty years invasive species have racked up a global bill of more than $2 trillion (USD).

The putative harms are more than just economic. South African authorities have calculated that invasive trees and shrubs around the drought-plagued city of Cape Town use more than one hundred million liters of water a day—water that's no longer going to recharge the aquifers upon which residents depend. Invasive mosquitos have caused new disease outbreaks among people in North America and Europe. On isolated islands, invasive predators upset the balance of ecosystems by literally eating the natives up. They are responsible for perhaps one in every three extinctions worldwide over the past five hundred years.

Concern about the issue knows no borders, but the United States has recorded more invasions than anywhere else, and it spends more than any other country to fight them. According to the US Fish and Wildlife Service, there may be as many as fifty thousand non-native species in the United States. Of those, 4,300 are officially considered "invasive," a definition that recognizes them as a threat to native ecosystems, human health, agriculture, infrastructure, or economic activity.

The US government estimates that invasive species cost the country $120 billion in environmental damage every year, and the federal budget earmarks some $3 billion to control them. About half that allocation goes to chemical controls (herbicides and pesticides). Other government projects include trapping, netting, bulldozing, and keeping invasive species from being introduced into new habitats with everything from electric fences to mussel-sniffing dogs.

The burden of invasive species control isn't just a problem for governments. The United Kingdom once conquered the globe, but now the British populace finds itself under siege by Japanese knotweed (*Reynoutria japonica*), a plant brought to London as a garden ornamental in 1850. In addition to being a persistent weed, knotweed can grow through concrete and infiltrate underground utilities with its deep roots. Today, Brits spend nearly a quarter billion pounds a year attempting to eradicate it, a goal that seems nearly impossible given that knotweed rhizomes can penetrate six feet deep and extend more than twenty feet horizontally, sending up new shoots along the way.

Reportedly, it only takes a piece of root the size of a human thumbnail to kick off a new knotweed invasion—and these scraps can survive underground for twenty years. Properties with any evidence of knotweed are notoriously difficult to sell, driving some homeowners to despair—or worse. In 2013, a fifty-two-year-old West Midlands man beat his wife to death and then killed himself, writing in a suicide note, "I believe I was not an evil man until the balance of my mind was disturbed by the fact that there is a patch of Japanese knotweed which has been growing over our boundary fence. . . . The worry of it migrating onto our garden and subsequently undermining the structure . . . has led to my growing madness."

Murder-suicide is fortunately an uncommon outcome of invasive species infestations, but run-of-the-mill madness is not. The reddit.com discussion group r/invasivespecies, for example, is rife with testimonials by frustrated gardeners battling knotweed, barberry, and Asian bittersweet (*Celastrus orbiculatus*), and Americans who despair at the abundance of European starlings (*Sturnus vulgaris*).

But diving into the matter, it turns out that which species we deem "invasive"—and what we try to do about them—often says as much about us as it does about the plants and animals we fear and loathe.

For one thing, humans love a good monster story—and the media loves to tell us one. In early 2022, viral news warned of an "invasion" of Joro spiders (*Trichonephila clavata*), "palm-size" arachnids from Japan that would, ecologists predicted, "descend" on the US East Coast that summer. The spiders, it turned out, are alarming in appearance but basically harmless. They might even play a beneficial role, as their prey includes insects like the spotted lanternfly (*Lycorma delicatula*), another "Asian invader" that is attacking the American tree nursery industry. But the words "giant spider" still proved irresistible, appealing as they do to our natural arachnophobia.

Snakeheads (family *Channidae*), large freshwater fish with Asian origins, are also apt villains, presenting a prime opportunity for horror journalism: "The snakehead is not a fish you especially want to meet when spending a peaceful day on the lake," warned a 2025 *New York Times* article, the latest entry in a very specific genre that dates back decades. "Aside from everything else, snakeheads—gulp—can breathe air. And they don't just look scary. Left unchecked, northern snakeheads can also threaten native species."

Not to put too fine a point on the matter, *National Geographic* has called the snakehead "Fishzilla."

Age of Anxiety

While many people just enjoy getting the shivers, stories about invasive species also tap into widely held feelings about environmental loss. In the 2020s, with the consequences of climate change—wildfires, droughts, storms—often dominating news cycles, social scientists began recording rising levels of so-called eco-anxiety, or even eco-grief. A recent survey of ten thousand people across ten countries, published in *The Lancet*, found the majority reported high levels of stress about the fate of the planet. Forty-five percent said that anxiety affected their daily lives, while 75 percent agreed with the statement, "The future is frightening."

Much of this has to do with predictions about change on a planetary scale, but anyone who has been on Earth for more than a couple decades also recognizes that the natural world we grew up with is, in many ways, irrevocably altered. In my state of Michigan, for example, winters bring far less snow than they used to. Some once common birds, like the evening grosbeak, are now a rare sight, while formerly ubiquitous monarch butterflies have become impossible to find. Lost biodiversity is nearly all attributable to human activity, but whether the direct cause in any one case is habitat destruction, overfishing, pollution, pesticides, the effects of a warming climate or some combination of all of these, there's often no neat way to assign blame.

But the story appears simpler—at first—when it comes to invasive plants and animals. The tale of a voracious snake or overbreeding rodent puts a face on ecological catastrophe, and hunting that enemy down lets us channel our anger, sadness, and fear into something that feels productive. (Such a campaign also won't force us to give up our own lifestyles or challenge powerful institutions—snakeheads, after all, don't employ lobbyists).

Thus, while the war on invasive species is led by governments and nonprofits, at a day-to-day level it draws in armies of regular people, who eagerly sign up to volunteer for nature: "Invasive species threaten every town's green space," warns the North American Invasive Species Management Association, "so it makes sense to involve citizen volunteers in invasive species removal."

Regular nature lovers most often contribute by pulling weeds in parks, forests, and nature areas. Sometimes, however, citizens are also deployed as literal bounty hunters. Witness "The Florida Python Challenge®," an annual ten-day event held in the Sunshine State during the sweltering height of August which offers cash prizes for slaying the most and longest Burmese pythons (*Python bivittatus*). Challenge organizers recommend only humane methods of execution, such as decapitation via machete.

Even as the headless snake corpses pile up, however, a prelapsarian version of the natural world remains frustratingly out of reach. Indeed, some ecologists and officials are quick to admit that public campaigns likely do little to stem the tide of invasive species. They argue instead that, like asking individuals to recycle plastics and switch to LED lightbulbs, organizing citizens to remove unwanted plants and animals may serve as a way to draw them into the broader cause of environmental stewardship, creating constituencies to demand that governments do more.

Killing animals, however, carries a different moral weight than ditching plastic drinking straws or writing letters to Congress. After all, it's not these species' fault that they find themselves so far from home, and an animal decapitated for the cause of environmental stewardship is not immune from suffering. What does it do to our souls when we're told the best way to help the environment is to destroy other living things?

In the sort of upper-middle-class community where residents have vegetable gardens and master's degrees, the topic of native and non-native species comes up surprisingly often when neighbors stop to chat. We're proud of our native wildflower gardens, disdainful of another's burning bush (*Euonymus alatus*; it comes from Asia), or upset that a pair of house sparrows (*Passer domesticus*, from Eurasia) have moved into our bluebird box. In rural towns, the conversation may be about Canada thistle (*Cirsium arvense*) in the pasture, or the best way to protect a cherry orchard from Japanese beetles (*Popillia japonica*). In the woods, it's all about garlic mustard.

The idea that some species belong here and others over there is central to the way we in the West think and talk about the natural world. Still, not every apparently clear-cut case of an invasive species survives a close, hard look. Sometimes the threat to native ecosystems turns out to be exaggerated, a species' bad reputation as much a product of hype as of science. Sometimes a non-native species brings with it benefits as well as harms. And sometimes the efforts required to remove an invader have their own detrimental impacts on the environment, or they raise troubling questions about how much violence we can—or should—stomach in the name of conservation.

Novel species can now be found everywhere in the world—even Antarctica, where several foreign arthropods and an earthworm have found hospitable ground along the continent's warmer fringes. But even those who prosecute the war on invasive species have trouble pinpointing exactly when everything started to go wrong: Was it when humans first arrived in a certain place, or when Europeans, specifically, arrived? Are the familiar plants and animals of Great Britain, many of which were imported by the Romans two millennia ago, native or not? Are America's feral horses (*Equus equus*) an invasive species, or were the Spanish conquistadors who brought them here merely reintroducing a group of animals that vanished from the continent eleven thousand short years ago? (Does it change anything if the native horses went extinct in part because humans hunted them?)

Exactly what we're hoping to put right is an open question, and today, in the age of climate change, it might be a moot one.

Climate scientists have predicted that by 2070 nearly 20 percent of the Earth's landmass will be too hot to sustain human life. This 20 percent includes areas that are, at present, highly populated, including almost the entire territories of India (population 1.46 billion), Indonesia (286 million), and Nigeria (237 million). Should this scenario come to pass, all of these climate refugees will have to move to cooler parts of the globe, where—if history is predictive—they will likely be seen as unwanted invaders.

Plants and animals are also migrating. According to a major paper recently published in the journal *Science*, global warming is accelerating changes in the distribution of Earth's plants, animals, and pathogens, as species shift their ranges to take advantage of milder conditions—or as their historic ranges become inhospitable. Can these species be stopped at our borders?

Among themselves, biologists, ecologists, and conservationists are already talking about whether they need to rethink their approach to dealing with invasive species, and even whether to revise the concept itself. But these conversations are relevant beyond the sciences, and not only because of the enormous budgets and energy that go into fighting this war. They also reveal much about ourselves, about our relationships with the natural world and each other, and about notions of home. Humans, who have spread across the globe in what is essentially an instant on the scale of geological time, have also displaced other species and disrupted ecosystems. Many a wit has noted that we may be the ultimate invasive species. And looked at that way, are we really that different from mute swans?

Toweling off after another summer day at the lake, this time with no swans in sight, but plenty of seagulls—and lots of humans—I decided to find out. The next time I got the call to step up and fight invasive species in my figurative backyard, I would do my patriotic duty and enlist.

I

INTO THE WOODS

On Thursday afternoons in spring, the Wildflower Rangers don boots, long pants, and gardening gloves to spend a few hours defending the forests around Ann Arbor, Michigan, from a spicy foreign enemy known as *Alliaria petiolata*—a.k.a. garlic mustard.

The name of the group is a sly bit of marketing, admits Matthew Bertrand, volunteer coordinator with Washtenaw County's Parks and Recreation Department, when I join him for a sortie at a nature preserve during an unseasonably warm May. Of course, he's asking us to pull weeds, the sort of tedious chore my fellow rangers and I—who skew upper-middle-class, educated, and suburban—tend to shirk when it comes to our own backyards. Calling us "Wildflower Rangers," on the other hand, lends the task an air of adventure, along with the positive feelings that come with defending some of our state's best-loved botanical treasures.

"Who knows what this is?" asks Bertrand, stopping to yank up an unassuming, leggy stalk from beside the trail.

"Garlic mustard!" some of us chirp, having read up in advance. (Washtenaw County, which hosts the University of Michigan's flagship campus, is populated by people who do their homework.)

In the biennial plant's second year, garlic mustard appears as a flowering stalk that can grow more than three feet tall (in its first-year form, as a tiny "basal rosette," it's far less distinctive). Bertrand points out the heart-shaped leaves, with their toothy margins, and the tiny, white, four-petaled flowers clustered at the top. If we had left this plant alone, he explains, it would soon develop skinny green seedpods known as siliques. A mature individual might carry several thousand tiny seeds, which, in the ominous phrasing of one ecologist, are "ballistically expelled" as many as six feet from the mother plant after the siliques ripen and dry out later in the summer. Any efforts to remove garlic mustard must take place before it reaches that phase—known as "seed shatter"—because once in the soil, some portion of the plant's miniscule progeny will remain viable for years, waiting for the right time to germinate and start the cycle all over again.

Uprooting garlic mustard isn't hard—a gentle, two-fingered tug usually does the trick. At first, the difficulty lies in spotting it among the many other green, leafy things that carpet the forest floor, competing for sunlight beneath the rapidly regenerating canopy of oak, maple, and hickory trees. Some of these things are poison ivy (*Toxicodendron radicans*, a native plant, and not our target), but as we're adequately kitted out, with our pants tucked into our socks to guard against ticks, Bertrand gives us the go ahead to venture off trail. Armed with heavy-duty yellow plastic bags, we're soon homing in on garlic mustard and yanking it out by the handful.

Herbal Colonialism

Though it looks more like a weed than a vegetable, garlic mustard is a member of the Brassicaceae family, a relative of turnips, cabbage, and kale with the same bitter kick as mustard greens. Like these farmer's market regulars, in its native range, which extends from the United Kingdom and Scandinavia to the foothills of the Himalayas, garlic mustard has a long history in the kitchen. In 2013, a team of archaeologists identified its seeds in food residue found on Neolithic potsherds in Germany and Denmark, the earliest evidence to date of Europeans using spices in their cuisine.

Europeans also favored garlic mustard for its reputed medicinal properties. According to herbologist Maud Grieve's exhaustive 1931 tome *A Modern Herbal*, the leaves could be eaten to induce sweating—assumed to clear toxins from the body—and applied topically to gangrene and ulcers. "The juice of the leaves taken alone or boiled into a syrup with honey is found serviceable in dropsy," Grieve added, before going on to note the garlic-scented herb's culinary value: "Country people at one time used the plant in sauces, with bread and butter, salted meat and with lettuce in salads. The herb, when eaten as a salad, warms the stomach and strengthens the digestive faculties."

Given its many uses, garlic mustard was most likely imported to the northeastern United States by European immigrants at least as early as the 19th century, accompanying other staples of European country kitchens and medicine cabinets that have since naturalized to North American parks, roadsides, and lawns.

Chickweed (*Stellaria media*) is either an annoying lawn weed or a "superfood" depending on how you look at it; early settlers valued it as one of the first edible greens to pop up in early spring. Common mullein (*Verbascum thapsus*), with its unmistakable six-foot flower stalks, arrived in the 18th century with Europeans who used it variously as a fish poison, anti-diarrheal, and respiratory medicine. Dandelions (genus *Taraxacum*), used to make wine, tea, and remedies for complaints ranging from gall stones to hemorrhoids, were reportedly packed on the *Mayflower*. Broadleaf plantain (*Plantago major*), a plant used in salads and topical salves that has been found in the stomachs of Iron Age bog bodies, earned the nickname "white man's foot" as British colonists spread it from New England to New Zealand. These herbs' hardiness and ubiquity in their native range likely contributed to the strong relationships Europeans formed with them, but such characteristics also enabled them to thrive as transplants halfway around the world.

However welcome it may once have been in a settler's garden, by the mid-20th century garlic mustard had made a thorough break with domestication. By 1950, land managers had spotted it in at least five US states and two Canadian provinces; forty years later, it was in eighteen states across the Midwest and Northeast. Today, garlic mustard's adopted range stretches from Maine to Alaska and across much of the South, and it regularly makes the hit lists of land management and conservation organizations nationwide, who run what are likely thousands of volunteer garlic mustard pulls every year.

"Hidden to the eye amongst this beautiful sea of green are plants intent on exterminating all others and gain[ing] control of the earth," warns the National Park Service in a call for volunteers at Effigy Mounds National Monument in Iowa. "Disguised as a delicious tasting flow[er]ing green plant, 'garlic mustard' is poised to dominate the plant life of the United States."

"[O]ne of the worst invaders of forests in the American Northeast and Midwest," warns New York State's official online "gateway to science-based invasive species information," on a webpage that recommends the use of the herbicide Roundup for stubborn patches.

"Michigan's worst woodland weed," declares a Michigan State University Extension report, urging Michiganders to "keep a vigilant eye and remove it as soon as it appears."

Indeed, the case against garlic mustard initially appears open and shut. While nearly seventy different insects consume *A. petiolata* in Eurasia, North American herbivores tend to avoid it—after all, it's spicy! Instead, they prefer to graze on the native forest plants that share garlic mustard's preferred habitat in the forest understory. Spared the pressures of predation and competition, the foreign herb is free to propagate itself, forming dense stands where native plants can't regain a foothold—an alarming situation to any land manager who comes across such a monoculture.

While it may have few enemies this side of the Atlantic, garlic mustard still produces the sophisticated chemical self-defenses it evolved under more difficult conditions in the Old World. Sulfur- and nitrogen-based glucosinolates, which also give the plant its pungent flavor, break down into compounds that are poisonous to fungi, insects, soil bacteria, and other plants (although not to humans, since our habit of chopping and cooking green vegetables destroys the toxins before we eat them). Garlic mustard, especially in its first year, seems to produce far more cyanide compounds than do native North American mustards—as much as 150 times, according to one study. In experiments, *A. petiolata*'s various allelopathic chemicals—compounds that harm neighboring species—have been shown to suppress both soil-dwelling microorganisms and the seed germination of native plants. (The phenomenon of allelopathy remains somewhat controversial, however, as its impact in the wild is hard to pin down.)

The mustard family's chemical defenses evolved as protection from predation, but they also serve as a positive signal to butterflies of the genus *Pieris*, which lay their eggs on brassica plants. This poses a problem for some North American species, for which garlic mustard could prove to be a "population sink." These butterflies have historically laid their eggs on North American members of the mustard family; eggs mistakenly laid on *A. petiolata* are less likely to hatch, and the caterpillars that do emerge are less likely to survive.

The invasive herb's habits are especially frustrating to conservation organizations because shade-loving garlic mustard can grow deep in mature forests. Unlike some introduced plants—including chickweed, mullein, dandelion, and plantain—that thrive mostly in turf lawns and disturbed areas, *A. petiolata* can encroach on what ecologists consider "high-quality habitat"—that is, a place that provides a diverse community of native species with everything they need to survive and reproduce. And given all the other threats facing these special places—development, pollution, climate change—a new stand of garlic mustard will often raise both hackles and alarm bells.

Katie Grzesiak, who now manages terrestrial invasive species for Michigan's Department of Natural Resources, will never forget one particular stand of garlic mustard discovered at Pictured Rocks National Lakeshore, a unit of the National Park Service in Michigan's Upper Peninsula. It was the plant's first appearance in this 114-square-mile preserve, where sandstone cliffs stained with multicolored mineral deposits tower up to two hundred feet over Lake Superior. While boat tours beneath the rocks are a popular tourist attraction, the park's interior remains a wilderness of sand dunes, old-growth cedar swamps, and upland northern hardwood forest that shelters such marquee species as gray wolves, black bears, and even the occasional moose. Though most of these woodlands are second-growth, the trees now average 100–130 years old and are largely unmolested by humans, who are shunted to rugged hiking trails and a few backwoods campsites. But this stand of garlic mustard wasn't even near a trail—it wasn't near anything. More than a decade later, Grzesiak still wonders how it got there.

"We had to hike a long way off trail to get to where that garlic mustard was," says Grzesiak, who led a seasonal exotic plant management team for the National Park Service for five years, beginning in 2008 when she was as an undergraduate at nearby Northern Michigan University. "It was found in one of these random surveys, in a pristine area. It was otherwise so beautiful! That was one of the reasons that it was such an important priority for us. We used to do some of this and do some of that, and instead it was like, 'No, we're going to spend a month pulling this stuff.' That completely changed the crew that I ran the whole spring."

The site's remoteness made for a challenge, but the crew was also stymied by topography and weather, as the garlic mustard had taken root on a sloughing pile of sediment below soft cliffs.

"There was a lot of space for those seeds to germinate and just keep going," Grzesiak says. "I spent, I think, three or four years working on that patch, and I definitely saw improvement, but it wasn't one of the most impressive improvements I've ever seen with garlic mustard. That one was . . . it was really tricky."

Grzesiak is the first to admit that not every site in Michigan is worthy of such an effort. In the state's southeast, which includes Detroit, Ann Arbor, and their suburbs, garlic mustard is now endemic.

"The farther north you go, the less of it there is, but I think it's in every county—if it's not, I'm shocked," Grzesiak says. "And in a lot of areas it's very, very thick. So from a statewide perspective, we aren't able to put as much of our time and effort into active management."

While the broader territory has long been ceded, garlic mustard remains a top priority for many smaller agencies and organizations, including Washtenaw County Parks and Rec. Michigan authorities, like their counterparts in other states, use grants and public communication channels to encourage local springtime pulls in certain high-value areas, such as the preserves targeted by the Wildflower Rangers program. Meanwhile, for Grzesiak herself, the issue is literally in her backyard. A few years ago, she and her husband moved to a former pine plantation outside Traverse City, and the couple are working to restore their ten acres of forest to a more natural state.

"In my own yard I think of garlic mustard as a pretty high priority," she says. "I've got a lot of red pines, but then growing in underneath them, we're getting some hardwoods. And that's where the garlic mustard really thrives—right where those little hardwoods are starting to come in."

Fighting for their share of sunlight, those little hardwood seedlings signify the resilience of a native ecosystem that humans once all but destroyed. When Europeans first set foot here, Michigan was so heavily forested that water routes offered the only practical means of travel. ("One cannot help fancying that he has gone to the ends of the earth, and beyond the boundaries appointed for the residence of man," wrote the geographer Henry Rowe Schoolcraft, who explored northern Michigan in 1820.)

The woods that to Schoolcraft had seemed dark, forbidding, and nearly endless were in fact easily vanquished. Between the 1850s and 1910, timber companies clear-cut the Lower Peninsula and much of the Upper Peninsula, sending lumber coast to coast for houses, railroad ties, and myriad other uses. What remained was a wasteland of stumps and brush—ready fuel for enormous wildfires that killed hundreds of people, wiping entire towns from the map.

It was not long before Michiganders began to have regrets.

Preventing fires became a statewide obsession in the 20th century, with everyone from township supervisors to the Boy Scouts taking up the cause. Railroad smoking cars were plastered with notices asking passengers to refrain from throwing match ends and cigarette butts out the windows of moving trains with the admonition, "Remember Metz and Au Sable," two towns destroyed by conflagrations in 1908 and 1911. "[T]he fires not only burn the growth on the lands, but burn up the land itself, its virility, making it little by little, a desert more and more incapable of reclamation," lamented the *Detroit News* in 1920.

As soon as wildfires were under control, people began working to recreate a semblance of the once-endless forest. During the Great Depression, some 125 Civilian Conservation Corps camps dotted the state, an echo of the long-gone lumberjack barracks, and "Roosevelt's Tree Army" planted vast tracts of fast-growing red pines in orderly grids, a pattern visible to this day even as other tree species have returned alongside them.

Restoring Michigan to its original state after so many decades of human exploitation was always going to be a challenge, however. Much of what was lost is gone forever, and the patches that remain only drive home the point. See, for example, Hartwick

Pines State Park, in Crawford County, where a scant forty-nine acres of old-growth evergreens, some three hundred years old and more than 150 feet tall, are nearly all that remains of the primordial woodland. This solitary, cathedral-like grove testifies to a Michigan that simply is no more.

Still, pulling garlic mustard when it emerges each spring is a simple, doable thing, Grzesiak argues. It's something responsible landowners can take on themselves, without any fancy tools or techniques. Meanwhile garlic mustard is a valuable "gateway species," she says, for educating everyday Michiganders on the impact invasives are having on their state.

"It's a great species for that," Grzesiak says. "It's not toxic to humans. It's pretty easy to pull. And you can see the difference. With some of our invasive species, you have to use herbicides or much larger efforts, but manually pulling this plant is a great strategy."

That's why each spring she will be outside uprooting garlic mustard, standing in spirit with volunteers like the Wildflower Rangers.

The Apostates

When Bernd Blossey and his wife, Victoria Nuzzo, first settled on a 300-acre patch of woods and meadow in central New York more than twenty years ago, they, too, pulled *A. petiolata* wherever they saw it.

"We spent every weekend and evening in spring pulling garlic mustard on this slope until we were sick and tired of it," Blossey says, when I meet the couple at their home one afternoon in July.

A professor in the Department of Natural Resources and the Environment at nearby Cornell University, Blossey, whose serious bearing belies a wry sense of humor, is leading me on a tour of the hilltop spread that doubles as the couple's home and research station. From behind their house, a bucolic view stretches across a meadow dotted with milkweed, toward rolling 1,800-foot ridges covered in mixed hardwood forest. Blossey and Nuzzo have aerial photos from the mid-20th century that show their property still in agriculture; today, the trees are between sixty and eighty years old: ash, oak, beech, and maple towering over a dense understory.

"We'd schlep bags of garlic mustard out of the woods," says Blossey, in a crisp German accent that has stuck with him over more than three decades in the United States. "And that was all that we did! We thought, 'This is ridiculous, because we need to do it over and over and over again! We would like to enjoy an afternoon together on the porch—something different than pulling garlic mustard.'"

"And when we pulled, the patch would be even bigger the next year," Nuzzo adds. Soft-spoken, with the air of someone who is far more comfortable outdoors than in, she recently retired as an independent consulting ecologist.

The persistent weed may have been a perennial frustration, but Blossey and Nuzzo could never bring themselves to hate garlic mustard—it is, after all, what brought them together.

"We're a garlic mustard couple, if you want to put it that way," Blossey says.

Growing up outside Hamburg in the postwar decades, Blossey originally wanted to be an arctic ecologist, exploring the mysteries of the polar regions like Hamburg's own Alfred Wegener. (A pioneering climatologist, Wegener is today best-known for his theory of continental drift, which was still met with skepticism when he died on an expedition in Greenland in 1930.) During his first year at Kiel University, however, Blossey discovered the possibilities of entomology—namely, that he could fill his room with aquariums and terrariums to observe nature up close, with no need to secure passage on an icebreaker.

"Fascination comes with the ability to observe," he explains.

In the 1980s, Blossey followed his evolving interests to a PhD project investigating biological controls for an invasive species that had begun to rankle American land managers. Purple loosestrife (*Lythrum salicaria*), a six-foot Eurasian perennial with showy magenta flowers, was allegedly disrupting water flow and crowding out native sedges, rushes, and cattails in wetlands throughout North America. (Imported both inadvertently in ships' ballast water and deliberately as an ornamental, by the 1980s loosestrife had a reputation as a "wicked weed" that was "steeping young American toads in poison," according to newspaper accounts at the time.)

Biological controls—in most cases, insects that feed on and damage a targeted invasive plant—have often been controversial, and any successful introduction requires years of field research and careful studies under quarantine to ensure the control will cause no harm to the ecosystem it's intended to save. When done right, however, the approach can be highly effective, and Blossey soon hit upon six promising European insects—three weevils, two beetles, and a gall midge, all of which lived solely on loosestrife and hindered its growth and reproduction. In 1992, he accompanied his bugs on their mission to the United States, a move that would prove permanent.

"I came with them initially just as a caretaker," Blossey says. "I would teach the wildlife biologists how to deal with the insects. The controls worked really well, and it all looked really good. And then the land managers around me said, 'Oh, you need to repeat this because one of the scourges in the woods is garlic mustard.' And I said, 'OK, let me look into that one.'"

Blossey wasn't familiar with garlic mustard, but as he pored over what research there was, he kept coming across Nuzzo's name.

"So I got a phone call out the blue," Nuzzo says with a laugh.

Nuzzo remembers with perfect clarity the first time she encountered *A. petoliata*: "Severson Dells, southwest of Rockford, Illinois, 1977, I saw the very first patch. I didn't know what it was."

After that, it was suddenly everywhere, spreading in conjunction with a decline in the diversity and abundance of native plants. Nuzzo, who consulted for state conservation agencies, would spend more than a decade observing these troubling changes in the upper Midwest. For a series of papers and reports that are still widely cited in the scientific literature today, she investigated garlic mustard's role in forest communities and experimented with methods of controlling it, from pulling to poisoning to burning.

But as Nuzzo returned to study plots where garlic mustard had once seemed poised to form daunting monocultures, a more complicated picture began to emerge. "In the woods that started with *some* herbaceous vegetation under the garlic mustard, when I went back years later the garlic mustard had simply crashed, and all the native stuff was still there coming up," she says.

When Nuzzo and Blossey began collaborating—at first, it was strictly professional—this apparent contradiction was something of a mystery. Coming off his success against purple loosestrife, Blossey, by then at Cornell, began a research program to investigate European insects that might help suppress *A. petiolata* in North America. But meanwhile, he and Nuzzo wanted to understand the puzzling dynamics at play in the years after garlic mustard gains a foothold.

"We were expecting garlic mustard to run through a system, because that's what people had explained to us," Blossey says. He sent grad students to the front lines, to see what happened when garlic mustard reached a new site and began expanding. "But it didn't expand," he says. "It declined. And other plants didn't respond to it in the way we thought. At some point it became clear that garlic mustard is not a driver, but probably a passenger in ecological change."

Garlic mustard had begun popping up at the same time that forest plants were coming under other new pressures. One was earthworms, the seemingly ubiquitous invertebrates that—like so many other invasive organisms—colonized North America alongside European settlers, who inadvertently transported them in their ships' dry ballast and in the root balls of horticultural specimens. Since the 1600s, earthworms of the family Lumbricidae had been gradually spreading west and north from East Coast population centers; in the 1970s, they were relative newcomers to the Midwest forests where Nuzzo worked.

Humans count on earthworms to aerate soil in our gardens, manufacture humus in the compost pile, and wriggle on the ends of our fishing lines. Meanwhile, an assortment of native birds—including American robins, which eat up to fourteen inches of earthworms a day—have happily adopted them as a dietary staple. But the familiar night-crawler is also changing the soil nutrient cycle in American forests in ways that make things difficult for many native plants.

North America once had its own earthworms, and still does in parts of the southern United States and along the Pacific Coast. In the Northeast and Midwest, however, they were extirpated by glaciers that scrubbed the continent during the last ice age. Today's wildflowers and ferns evolved in their absence, adapting to a worm-free woodland.

Without earthworms, the thick layer of dead leaves and other detritus—known as "duff"—on the forest floor is gradually decomposed by native microbes, fungi, and soil-dwelling invertebrates like nematodes and millipedes. This slow decay provides a steady source of nutrients for tree seedlings and other shallow-rooted plants, while lingering top layers insulate the rhizomes and tubers of spring ephemerals from winter cold and act like a sponge to maintain ideal moisture levels.

Earthworms, however, digest duff more quickly than native plants prefer. They also dig deeper than native soil-dwellers, mixing nutrients into lower layers where they're only accessible to deeply rooted trees and can be leached out of the forest altogether by groundwater. These dramatic changes may harm native plants, but they do not pose a problem for garlic mustard, which evolved in Eurasian forests where earthworms have always been a factor.

"I used to have a $5,000 reward if somebody would come and give me a location where garlic mustard goes, and there are no worms," Blossey says.

No one ever claimed the prize. And meanwhile, a very different animal's growing population was upsetting the balance above ground.

The oldest extant deer species in the world, whitetails (*Odocoileus virginianus*) are native to the Americas, where they originally ranged as far west as the Rocky Mountains and were a staple food of both Indigenous people and large predators, such as wolves and cougars. With the arrival of Europeans, however, deer hunting became an industry; Native Americans increased their whitetail harvest to trade for European goods, and settlers cleared habitat while shooting vast numbers of deer themselves. In the early 19th century, organized groups of market hunters sometimes took down hundreds in a day, but by the dawn of the

20th, deer were gone from more than a dozen states. Isolated populations found refuge at the extremes of their range, including in the Adirondack Mountains and northern Maine, and in 1890 the US Biological Survey estimated the total at only about three hundred thousand whitetails nationwide.

Today, that number stands at more than thirty million.

White-tailed deer's remarkable rebound represents a triumph of the early conservation movement, which worked to protect the animals from overhunting and to restore their habitat. But deer have also benefited incidentally from other changes in how Americans use land and relate to nature.

In the 20th century, second-growth forest succeeded timberland and pastures, providing a greater density of saplings—and thus better browsing—than pre-settlement old growth ever did. That has allowed deer to thrive in areas historically less hospitable to large herbivores. Meanwhile, deer adapted happily to the patchwork of woods and lawns in suburbs and exurbs, places where hunting is prohibited and predators nonexistent.

And all these deer have voracious appetites. Adults, topping out between 150 and 200 pounds, must eat the equivalent of 6–8 percent of their body weight every day between spring and fall. But along with their immense caloric needs, they are picky about nutrient content and digestibility. In the spring, about a third of their diet consists of tender forbs, the botanical term for herbaceous flowering plants that includes spring ephemerals like trillium (which, in fact, they consider a special delicacy).

Deer do not, on the other hand, eat garlic mustard.

"The understory is to some extent determined by how many deer you have," Blossey says. "And the pattern that deer are favoring invasive plants—that is, not eating them—is pretty generalizable."

As Blossey and Nuzzo studied garlic mustard in its ecological context, the role of worms and deer in *A. petiolata*'s colonization of North America became apparent, complicating the seemingly simple story of an invasive species running rampant through a naïve ecosystem.

In the 1990s, Nuzzo worked as a consultant for Fermilab, the massive US Department of Energy facility outside Chicago. Officials there had been watching with alarm as the population of white-tailed deer exploded on their 6,800-acre campus, causing traffic accidents and environmental degradation. The lab, home to a massive particle accelerator (along with a small herd of American plains bison), had recently restored one thousand acres of forest, wetland, and tallgrass prairie, but those efforts attracted deer—as many as fifty per square mile according to one estimate, or ten times the assumed ecological carrying capacity for the species.

At Fermilab, Nuzzo fenced off seven "exclosures" to keep deer out and compared these over five years to open control plots. The difference was stark: plant life thrived within the fenced areas and declined precipitously where deer still browsed, in some places leaving nothing but bare earth. In part as a result of Nuzzo's work, the US Department of Agriculture recommended that Fermilab cull deer by employing sharpshooters. By 2005, these USDA-trained hunters were maintaining the deer population at a tenth of its peak, and forest groundcover had increased from 30 percent to more than 80 percent. Meanwhile, Nuzzo, in a follow-up study conducted with Blossey, found that garlic mustard steadily declined once the deer were under control.

"They're nice, they're pretty, but they can be a real problem," Rod Walton, a Fermilab ecologist, told the lab's employee newsletter in 2006. "They were literally eating us out of house and home."

As Nuzzo and Blossey worked on the deer and earthworm questions, they also discovered something surprising about garlic mustard itself. When allowed to take its course, *A. petiolata* seems to keep itself in check through a process called "negative soil feedback."

"Part of it was driven by observations that we had here," Blossey says, as we walk past the couple's vegetable garden (like Nuzzo's exclosures at Fermilab, it is surrounded by a high fence—deer, after all, are everywhere).

Some of the oldest patches of garlic mustard the couple had observed on their acreage became short and sparse over time without intervention, they noticed. Meanwhile, in more recently colonized areas, the plants were hip high. It's a puzzling but common pattern, which the couple's experiments suggest is caused by a process that may be better recognized in agriculture.

"You know you shouldn't be planting potatoes after potatoes after potatoes," Blossey says. "You need to have some rotation. And the reason for that is that the soil builds up microorganisms that would suppress that particular plant—maybe fungi, maybe bacteria, maybe other things. And so garlic mustard does the same thing, almost all plants do. The longer it sits in the same piece of soil, the more it will be suppressed. Its performance will be reduced."

With this fact comes an irony that calls into question all the work my fellow Wildflower Rangers and I were doing back in May.

"If you go in and yank that out," Blossey says, "you take the food for the microbes away. So the next time a seed germinates—boom, it can grow again to the same height. When you're weeding, it actually backfires, because you set back that self-poisoning."

For Blossey and Nuzzo, the lessons were clear. Blossey took up deer hunting on his property, and trilliums and oak seedlings started to return. Meanwhile, the couple stopped pulling up garlic mustard, and they were rewarded with less and less of it.

They are explaining this as we walk down their long driveway, to where they remembered seeing some garlic mustard earlier in the year. By mid-summer, the plant is in its seed shatter stage, and when we spot the handful of stalks beside the drive, they're dry and gray, a nearly unrecognizable shadow of the leafy green herbs I was yanking up back in the spring. I reach out to touch one of the shriveled siliques and am startled by a small explosion of tiny black seeds, which land invisibly among the leaf litter: I've just helped an invasive plant propagate itself. Having internalized the messaging of Michigan's DNR and the Wildflower Rangers, I feel a twinge of guilt. But Blossey and Nuzzo are not alarmed.

"If we make a recommendation to intervene, we should be improving the condition for the things that we care about," Blossey says. When that improvement doesn't follow, he explains, scientists need to rethink their message. "We need to be able to say, 'I have new evidence, so I'm going to change my recommendations.'"

Following the couple's lead, scientists have gradually embraced nuance when it comes to *A. petiolata.* Over the past decade, other researchers have found that while some native plants may do worse in the presence of garlic mustard, it correlates with better growth and survival for others. Meanwhile, researchers have observed surprisingly fast evolutionary adaptation in some native species that now share their ranges with garlic mustard. Two decades ago, for example, garlic mustard was known to be lethal to most caterpillars of the native green-veined white butterfly, *Pieris oleracea,* likely contributing to its decline. But recent research has shown that over time *A. petiolata* has successfully been incorporated into the larval diet—and the butterflies' numbers are now on the rise.

Still, this message has been slower to reach those who confront garlic mustard on the ground.

"We were able to back it up with data, and then still people were not necessarily believing us," Blossey says. "The land managers didn't like that it was questioning what they had seen. But I can't control what people think, what people do. I can only provide some evidence."

Paying the Rent

There's the sort of evidence you might read about in scientific papers, and then there's the evidence you can hold in your hand—or the kind that piles up, in a big stack of yellow plastic bags stuffed with garlic mustard. The evidence of a productive day's work.

"I definitely feel like I need to be of service, especially at my age," says Patty Lorandos, a sixty-eight-year-old grandmother who joins the Wildflower Rangers for a garlic mustard pull at Whitmore Lake Preserve, half an hour from Ann Arbor, one sunny May afternoon. Established in 2013, the preserve had previous lives as a vineyard and, later, a venue for multi-day music festivals. Signs of the latter are still easy to see—a gently sloping meadow retains the shape of an amphitheater, while moldy thermos bottles and other discarded camping equipment turn up from time to time among the poison ivy in a grove of oaks nearby.

A self-identified "counter-culturist," and the sort of free spirit who once attended such rock and roll festivals herself, Lorandos laments that her generation is responsible for irremediable ecological damage. "Supposedly we had all the answers and we were going to make things so much better," she says. "I look in horror at what younger generations must contend with. But I absolutely feel we should never give up, never surrender."

To that end, pulling garlic mustard is just one of the volunteer efforts to which Lorandos, a former teacher and office administrator, has devoted her retirement. In election years, she knocks on doors for Democratic candidates (although "I don't believe in political parties," she says), and she signed up with the county to monitor vernal pools, the temporary spring ponds that have been described as the coral reefs of northern forests for the biodiversity they support. Like most people who have given it any thought, she knows garlic mustard is here to stay in Michigan, and she has even come to appreciate it as a zesty ingredient in salads. But removing *A. petiolata* from high-value woodland habitats feels like a way to preserve some semblance of an environment that has become hard to recognize in just the time since she was a young woman.

When Lorandos first moved to her home on the Huron River some thirty years ago, she says, "the level of birdsong was thunderous. And now that birdsong has faded. We had nesting birds that would return every year, and certain species that were thick around the house. I barely see them now."

Chatting amiably along the way, our group has hiked past the old campground and into a section of mature forest in the farther reaches of the preserve, where violets, wild geraniums, meadow rue, and trillium carpet the forest floor between maples and hickories.

"This work is about trying to save high value areas in the hopes that they can be a time capsule for biodiversity," Lorandos says. It's also good for the soul, she adds. "I mean, I definitely feel better after pulling weeds in the woods than I do going door to door and trying to get out the vote."

A similar feeling drives John Metzler, sixty-five, who sees volunteering with the county's Parks and Recreation department as a way to "pay the rent" on the nature preserves where he spends so much of his time hiking, taking photos, and getting back in touch with a natural world that he had stopped noticing over the decades he focused on working and raising a family.

I meet Metzler, a retired economist, on a day when the contingent of Rangers is unusually large. A dozen engineers have taken the afternoon off from their jobs at Ford Motor Company, in nearby Dearborn, for a company-sanctioned volunteer day at Rolling Hills County Park, a 439-acre spread that, in addition to wooded areas, includes ball fields, picnic grounds, cross-country ski trails, and a popular water park. As a more seasoned volunteer, Metzler wears hiking boots, khaki work pants, and a wide-brimmed sun hat. It's a look that presents a striking contrast with the white-collar Ford employees, whose baseball caps, ankle socks, and anxiously applied bug spray betray them as relative woodland neophytes. On the way to our work site, Metzler spots a lime-sized oak gall and offers the group an impromptu talk about the bizarre structure, the home of a tiny wasp larva that created it by chemically derailing the growth of the plant's tissue.

Metzler's knowledge of the natural world is impressive—he can name seemingly every species in the undergrowth—but when it comes to pulling garlic mustard, he takes a certain pleasure in not considering too deeply what it means for a species to be undesirable, or whether his efforts will actually improve the diversity of native plants growing in the woods. It's too easy to overthink this, he says—and once you do, it's hard not to also think about the fossil fuel we volunteers burned to get to this park, or all the other environmental sins humans commit just by moving through the world.

"The nice thing with the garlic mustard pull is it's not my job to decide what to pull, what not to," he says. "I can just be ignorant and feel like I'm doing good. I'm muscle, I'm not brain."

Matthew Bertrand, the Parks and Rec volunteer coordinator who conceived of Wildflower Rangers, is glad to see us enjoying ourselves. He has been tackling invasive species on and off for nearly two decades, and he has found that stoking fear about environmental threats is more likely to lead to burnout than to effective action.

"People care about nature, and this is using your body to destroy things, which is generally a pretty satisfying thing to do," he says. "'Invasive species control' sounds horrible, but this is fun."

And it *is* fun. Driving home past corn fields and suburban sprawl, I pick bedstraw out of my socks, into which my pants are still tucked. Metzler identified this plant for me; a native annual, it's called bedstraw because it was once used to stuff mattresses, but today it's better known for its ability to stick to almost anything (one of several members of the genus *Galium*, farmers call it "catchweed"). I tuck this fact away, fulfilling in some small way what seems an ancient human urge to recognize plants, to know when they bloom and bear seed, and whether they're useful or good to eat.

Whether we did any good for nature today is a thornier question, one that Blossey is happy to answer with a deflating "no."

"You're out in the woods, you're looking at wildflowers, it's a nice spring day," he says. "The problem, though, is you need to go back over and over and over again. Every year you go to the same places and you do it again. What nobody is actually thinking about is, is this actually helping the things that we want to protect? Or are you uprooting or stepping on little trillium seedlings?"

It's challenging to convince anyone his efforts have been in vain, especially when an activity makes such intuitive sense, Blossey notes. Sometimes, people have to see the evidence for themselves.

But that will take more than a few afternoons in the woods.

2

A BRIEF HISTORY OF INVASION

In his youth, Charles Darwin spent vacations at Maer Hall, his uncle's ivy-covered 17th-century manor house in England's West Midlands. The estate's extensive grounds allowed the budding scientist free range to pursue his interest in the natural world—"ample means of investigation," as he would put it later in *On the Origin of Species*—and to observe the dramatic impact human intervention could have on it.

Surrounding the estate, he recalled in *Origin*, "there was a large and extremely barren heath, which had never been touched by the hand of man." Set apart, however, were several hundred acres that had been fenced and planted with Scotch fir some twenty-five years previously. "The change," Darwin wrote, "was most remarkable."

Scotch fir, also known as *Pinus sylvestris* or Baltic pine, grows in northern and alpine regions of Europe and Russia. It was once found throughout Great Britain, but as the climate warmed after the last ice age the evergreen's range contracted. Prehistoric humans assisted that decline as they cleared forest to create pasture for sheep and cattle, which in turn munched on any seedlings that managed to sprout, effectively extirpating the species south of Scotland. In the 18th and 19th centuries, however, wealthy landowners had begun to reintroduce Scotch fir here and there on country estates, where it had a transformative effect on the immediate ecosystem.

Compared to the surrounding heath, a low shrubland characterized by relatively infertile soil, the fir plantation at Maer Hall was buzzing with life. Along with the trees, which had been planted by Darwin's relatives, came a dozen new wild plant species and six insect-eating birds that were rarely seen elsewhere in that part of the country. "Here we see how potent has been the effect of the introduction of a single tree," he wrote in *Origin*, "nothing whatever else having been done, with the exception that the land had been enclosed so that cattle could not enter."

It was an early lesson in "how complex and unexpected are the checks and relations between organic beings," Darwin wrote in a chapter meditating on nature's apparent harmony, which he saw instead as merely a stalemate in a great "struggle for existence." But Darwin also recognized how easily human intervention can tip that balance. At any given time, plenty of species are poised to take advantage of our ability to transport them into new environments, remove their enemies, or otherwise alter their habitats for better or worse.

Even before we arrived on the scene, however, nature's balance was never quite stable, and there was always something waiting to seize the opportunity of a good cataclysm. This tendency of some species to flourish after disruption while others decline and vanish is a longstanding theme of natural history. Two centuries ago, Darwin observed that pattern in the English countryside. Today, modern paleontologists are tracing it through the fossil record, deep into the past.

When Invertebrates Attack

"When I got started working on invasions, it was for my PhD work, right at the turn of the century," says Alycia Stigall, adding after a beat, "I like to say that because it's funny." Her fellow Gen Xers, after all, still consider the "turn of the century" to be 1900, while her current crop of undergraduates wasn't even born at the most recent turn.

Of course, Stigall herself operates on a whole other timeline.

A professor of earth and planetary sciences at the University of Tennessee, Knoxville, she works out of a sunny office lined with shelves of fossils, many of which she or her students chipped out of roadside cliffs around the lower Midwest. These artifacts help tell the story of the Richmondian Invasion, one of the most momentous in history—at least, from the perspective of a brachiopod.

"The environment that you want to imagine is like the one off the coast of Florida," Stigall says, evoking the warm marine ecosystem that overlay what is now Greater Cincinnati, Ohio, roughly 450 million years ago. "Nowadays you would see corals and clams, but back in the Ordovician Period, you would see a bunch of things that look like clams, like this"—she picks up a golf ball–sized fossil—"but they're not."

Brachiopods, the dominant creatures in this shallow sea, were not related to modern mollusks at all. Instead, they formed a distinct phylum, one which today is extinct except for a handful of rarely seen holdouts haunting the deepest, coldest ocean waters. But despite their distance on the evolutionary tree, these ancient shelled creatures lived lives very similar to those of modern clams and mussels: as larvae, they drifted through the water column before settling on the sea floor, where they used a stalk-like "foot" to anchor themselves to something firm. They would remain in that place for the rest of their lives.

"These things don't move—the adults aren't doing anything," Stigall emphasizes. "They reproduce by throwing gametes into the water, hoping they get lucky and find the counterpart and produce an embryo. That juvenile floats around for a couple of weeks, and the currents will take it wherever it goes."

The Midwestern brachiopods carried on this lifestyle for eons, until the water level in their shallow sea began to rise. Within a few thousand years—a blip in geological time—their isolated basin merged with others, allowing larvae from elsewhere to find their way in and establish themselves. Among the invaders were new, different brachiopods, which competed with native species for habitat. They were accompanied by corals and bryozoans, colonial invertebrates that had never before existed in the Midwest basin.

"Now, corals, of course, are predators, so they're out there with their stinging cells grabbing plankton out of the water column," Stigall says, miming coral tentacles with the fingers of her left hand. "So new species settled on the sea floor, and some species did really, really well. And some of the species that *had* been really prominent previously went extinct."

The new pressures of predation and competition proved too much for the most specialized brachiopods, such as those that could only attach to certain types of sea floor—a resource that now may have been much more in demand. More flexible creatures, on the other hand, adapted to the upheaval, in some cases by changing their preference to a habitat that invading species did not occupy.

The Richmondian Invasion, notable in part because of the detailed fossil record that illustrates it, seems to be typical of prehistoric invasions. Environmental change is common over geological time, Stigall notes, but not every rise in temperatures or sea levels brings with it a transfer of plants and animals. That means that scientists examining the fossil record can compare times when environmental conditions changed with and without a concurrent movement of organisms, allowing them to isolate the impact of invasion on biodiversity.

It turns out that, while invasion in the short term may be bad news for biodiversity, over time, new arrivals drive the evolution of brand-new species as organisms shift their strategies and focus energy on resources for which there's less competition. When species adjust their ecological niche, they change over time to better fit it, and eventually a diverse array of entirely new species arises. Richmondian fossils, for example, show that brachiopods that survived the invasion evolved to be more specialized than their generalist ancestors. Still, settling into narrower niches can make descendants of both the invaders and the original native species more vulnerable during the next invasion. It's a cycle that Stigall and her colleagues have seen again and again.

"When the invasion starts, we lose the very specialized species," she says. "There is this interval that happens where there are just these generalists figuring out how to build a new world together, and they narrow their niches to partition that ecospace. And then afterwards, after that stabilizes, then we get back to producing new species."

The pattern can be roughly approximated by a sine curve, as biodiversity goes up and down with each invasion, whether it's a flood of strange invertebrates, as in the Richmondian Invasion, or the formation (via volcanic action) of the isthmus of Panama around 3.5 million years ago, which led to what's known as the Great American Biotic Interchange.

"That was an invasion, and it's one of everybody's favorites to talk about, because it's so recent," Stigall says.

It's also how the United States got its only marsupial.

Bridging the Gap

Though its origins might lie elsewhere, the Virginia opossum (*Didelphis virginiana*) long ago entered the idiom of the United States thanks to its unique habit of mimicking its own demise when threatened: "playing possum." Opossums also made their way into the cuisine of various human communities, including the Powhatan people of Virginia. They named the creature *aposoum*, or "white animal," and introduced it to their neighbors, the Jamestown colonists, who compared the creature's flavor to that of pork.

"An opussum is a beast as big as a pretty beagle," reported the 17th-century colonial chronicler William Strachey, with "a head like a swyne; eares, feet, and tayle like a ratt; she carries her young ones under her belly, in a piece of her owne skyn, like as in a bagg, which she can open and shutt, to lett them out or take them in, as she pleaseth, and doth therein lodge, carry, and suckle her young. . ."

Opossums were the first marsupials Europeans ever encountered, and Strachey was not the only one to marvel at their strangeness. At one time, however, pouched mammals were more widespread. Scientists now believe marsupials, whose premature young finish developing in their mother's pouch, actually first evolved in the Americas in the Jurassic period before spreading through the rest of the world. Not long afterward, however, they became extinct in North America, Africa, and Eurasia. Today, about 30 percent of marsupial species live in South America, while nearly all the rest are found in Australasia, where their ancestors migrated during a period when South America and Australia were connected by land bridges via Antarctica.

After the Isthmus of Panama formed, marsupials got another chance in the Northern Hemisphere, and at least some opossum species moved north, recolonizing an area that had long been the sole domain of placental mammals. The Virginia opossum made its way through Mexico to the southern United States, where it has thrived ever since.

Opossums were not the only creatures to venture across the isthmus. Fossils show that at least one species of the so-called terror birds, a family of flightless carnivores that may have stood ten feet tall, made it as far as Florida. The giant ground sloth, a character recognizable from Ice Age dioramas at regional museums across the United States, also traveled north. So did the armadillo, a south-of-the-border import that Texas has adopted as an official state symbol.

Despite a few marquee examples, however, North American migrants turned out to be far more successful in South America than those who traveled the other way. The Great American Biotic Interchange was bidirectional but ultimately asymmetrical—so much so that more than half of the mammal species living in South America today are descended from North American taxa, including some that seem to us quintessentially South American.

"You think of llamas and you immediately think of the Andes and Peru," says Juan D. Carrillo, a postdoctoral researcher in paleo-biology at France's National Museum of Natural History who researches the fossil record of the Great American Biotic Interchange. "Llamas are an iconic South American animal, an important animal for many native communities. But llamas are camelids, so they're related with camels, and camels actually originated in North America."

Carrillo grew up in Bogotá, Colombia, and studied biology at that city's National University, where he became fascinated by how fossils could reveal intermediate morphology—that is, the ways in which animals changed over time, one species evolving into another. Carrillo followed that interest to an internship at the Smithsonian Tropical Research Institute in Panama, which, although the dominant modes of travel may have changed in the last three million years, is still a busy hub of international traffic.

"One of my first jobs was to go to the canal and collect fossils," Carrillo says. "They were expanding the canal, so we would go every day and look, because after a while, they had to cover some of the sites with cement or flood some things."

More recent expeditions have taken Carrillo back to Colombia, where La Venta, a rich vein of fossils dating to more than twelve million years ago, offers the clearest picture of life in South America before the Great American Biotic interchange. While today the area is part of La Tatacoa Desert, it was once a tropical forest, home to a wide variety of now-extinct vertebrates, including rhinoceros-like ungulates and glyptodonts, massive cousins of today's armadillos. Carrillo's research shows that these native South American megafauna were already in decline as Panama formed, which may help explain why North American mammals did so well once they crossed the isthmus.

"We don't know exactly what could have caused the extinctions," Carrillo says, "but some possibilities are that this was related to environmental changes that happened at the time, that the climate was becoming drier and colder."

Carrillo notes that the dominance of mammals with North American origins is more pronounced in the continent's temperate plains and highlands than in its tropical forests. Such habitat would have felt familiar to North American mammals, who may have reached it by following the Andes like a sort of prehistoric Pan-American Highway. Perhaps North American invaders were taking advantage of disruption caused by climate change, filling ecological niches that had already been vacated. And perhaps they also brought new pressures—competition or predation—that helped drive some South American species the rest of the way to extinction.

Larger currents of evolution may also have played a role. For most of the fifty million years or so before the land bridge fully formed, South America was subject to what the paleontologist George Gaylord Simpson famously termed "splendid isolation": when its last connections with Antarctica and Australia were severed, South America became an island continent, one that bred peculiar mammals like *Thylacosmilus*, a genus of marsupial-like predators whose strangely shaped lower jaws sheathed knifelike canine teeth, and *Macrauchenia*, which looked like a cross between a horse and an elephant (its fossils famously baffled Darwin).

North America, on the other hand, still had intermittent land connections to Eurasia during this time, and, via Eurasia, to Africa, meaning that its animals theoretically evolved in competition with the greater number of terrestrial species that inhabited this larger portion of the earth's land mass. That they would be more successful in South America than the other way around is consistent with Darwin's predictions. As he wrote in *Origin*:

> As natural selection acts by competition, it adapts the inhabitants of each country only in relation to the degree of perfection of their associates; so that we need feel no surprise at the inhabitants of any one country ... being beaten and supplanted by the naturalised productions from another land.

It makes sense, then, that the cougar, *Puma concolor*, an apex predator and consummate generalist with evolutionary origins in Asia some eleven million years ago, now ranges comfortably from the boreal forest of Canada to the desert of the Southwest, through the Amazon, and all the way to Patagonia, while *Thylacosmilus*, already on its way out when the Panamanian isthmus formed, is now merely an oddity in the fossil record. Despite human persecution shrinking their territory considerably, cougars can thrive in many habitats while eating anything from deer and cattle to insects and mice. Although its diet is not well understood, the smaller, stranger *Thylacosmilus* seems to have been less able to adapt.

Today, the Panamanian land bridge is still the site of species exchange, only now human rather than volcanic activity is facilitating migration. Coyotes, native to the central plains of North America, are well suited to living near us. Thanks in part to the extirpation of wolves throughout most of the continent, these smaller, more adaptable canids spent the 20th century expanding their range from coast to coast. Now, they are moving south; in the 1960s, they made it to Nicaragua and Costa Rica, and by 2016 they were spotted for the first time on the other side of the Panama Canal.

"They do quite well with people," Carrillo says. "They're usually associated with agricultural lands, as well as clearing of forests. And this is unfortunately happening more and more in Panama."

According to recent research by scientists at the Smithsonian Tropical Research Institute, a population of coyotes is growing in the agricultural region north of the country's rugged Darién Gap, where they have taken advantage of deforestation and other anthropogenic changes. Whether they make it all the way to South America, and how well they do if they get there, will also depend on us. The thick rainforest of the Darién has never been penetrated by a real road, but surging human migration in the 2020s brought environmental disruption in the form of trash, trails, and felled trees—just the sort of situation a plucky generalist can take advantage of.

Here, as in much of the world, the future likely belongs to organisms that, wherever their origins, now live comfortably alongside people. Just ask opossums, for whom we have been nothing but a boon (except for the occasional, unfortunate individual that may find itself boiling in a stew pot or run over by a pickup truck). First, we wiped out many of the opossum's predators, while generating plenty of trash for them to eat. Then, settlers deliberately carried opossums north to Michigan and Maine and west to the Pacific Coast, where despite their non-native background, they seem to have found a crack public relations team. No longer just "good eating," the opossum has—in the words of one Malibu wildlife rehabilitation center—"a pivotal role in the ecosystem" as "nature's double-duty pest control and sanitation." And thanks to this mutually beneficial relationship, the species has claimed more space in a century than it managed on its own over the previous three million years.

From a Sinking Ship

The introduction of new organisms into new territories has long been one driver of evolutionary change. But it used to take place on a geological timeline, with the collision of tectonic plates and the rise and fall of seas and mountain ranges. Now, if they hitch a ride with us, plants, animals, and microbes can travel around the world in months, days, or hours. Among the many species we've carried with us are those that, like the opossum, create little conflict in their new homes. Others, however, prove less welcome from the very beginning.

Foremost among these is the rat, a creature whose range is now identical to our own, and whose presence we have historically associated with filth and pestilence. Although humans have lived alongside rodents since before the dawn of agriculture, it was only with the advent of transoceanic shipping that the mighty rat accomplished true world domination.

"Rats kept humans company, living, multiplying, gnawing and causing havoc on the ship, and yet taking full advantage of the human communication networks that they endangered," writes animal studies scholar Kaori Nagai in the 2023 edited volume *Maritime Animals*, a consideration of the creatures that accompanied seafaring humans—as cargo or stowaways—during the Age of Exploration.

As European countries expanded their colonial empires, rats came along, taking up their longstanding roles as urban and agricultural pests in places where they often posed a threat to local fauna that had evolved in their absence. Their spread did not go unnoticed by the humans with whom they hitched a ride—indeed, Europeans saw the fecund rodents as "master colonizers," Nagai writes. She quotes an 1857 article from the London magazine *Quarterly Review*:

> Scarcely a ship leaves a port for a distant voyage but it takes its complement of rats as regularly as the passengers, and in this manner the destructive little animal has not only distributed himself over the entire globe, but, like an enterprising traveler,

> continually passes from one country to another. . . . In this manner, many a hoary old wanderer has circumnavigated the globe oftener than Captain Cook, and set his paws on twenty different shores.

These shipboard populations consisted of two kinds of rats, the black rat (*Rattus rattus*), which has likely existed in Europe since at least the Roman era and was recorded in South America by the mid-1500s, and the larger, fiercer brown rat (*Rattus norvegicus*). The latter originated in East Asia but established itself in Europe in the early 1700s, first displacing the continent's native black rat and then spreading from European ports to the New World and the Pacific.

It was in the Pacific where rats as an invasive species would have the greatest impact on native ecosystems. Just one brief visit by a ship restocking its water supply was enough for brown rats to establish themselves on an island, often to the detriment of the locals. Eager omnivores, rats are particularly fond of birds' eggs and young, and they can quickly eliminate a breeding population—especially in places where birds evolved without terrestrial predators and lack any defenses against them. Tiny Hawadax Island, off Alaska, for example, became known as "Rat Island" once the rodents from a 1780 shipwreck had eaten their way through the native seabirds. The birds' disappearance reverberated through the food chain as populations of marine snails and limpets that had made up their diet exploded. The molluscs in turn churned through Hawadax's offshore kelp forests and drove off myriad creatures that had once sheltered there.

"Certain invasive species can have impacts beyond those that are most obvious," Carolyn Kurle, a conservation ecologist at the University of California, San Diego, told *Scientific American* in a 2021 article about a decade-long project to exterminate Hawadax's rats by air-dropping poison pellets.

To modern biologists, cases like "Rat Island" are tragedies, but to Darwin and his contemporaries the success of invasive species in isolated places was an interesting example of natural selection at work. "In many islands the native productions are nearly equaled or even outnumbered by the naturalized," Darwin noted in *Origin*, "and if the natives have not been actually exterminated, their numbers have been greatly reduced."

The obvious conclusion—one consistent with the later findings of scientists studying the Great American Biotic Interchange—was that successful new arrivals were better fitted for survival, having previously evolved in places with fiercer competition than is typically found on islands. "The species of all kinds which inhabit oceanic islands are few in number compared with those on equal continental areas," Darwin wrote, arguing that successful invasions prove that a "sufficient number of the best adapted plants and animals have not been created on oceanic islands; for man has unintentionally stocked them from various sources far more fully and perfectly than has nature."

In the minds of Victorian scientists, these non-natives' comparative "perfection" was all too easily extended to humans, especially at a time when Europeans, aided by gunpowder and disease, were conquering the far-flung peoples of the earth. "When civilized nations come into contact with barbarians the struggle is short," Darwin wrote in his 1871 follow-up *The Descent of Man*, as his fellow Englishmen enacted this grim narrative on a worldwide scale.

Indigenous people themselves, as Nagai points out, were well aware of the danger posed by new arrivals, whether they came on four legs or two. She cites a circa-1864 letter to Darwin, in which the German-born New Zealand geologist Julius von Haast marveled at "the changes which have taken place since first Captain Cook set foot in New Zealand." The brown rat, von Haast reported,

> is to be found everywhere, even in the very heart of the Alps, growing to a very large size. . . . The native (Maori) saying is, 'as the white man's rat has driven away the native rat, so the European fly drives away our own, and the clover kills our fern,' so will the Maoris disappear before the white man himself.

The connection between European colonization and the spread of plants and animals was never in dispute, but a century after the *Quarterly Review* writer's "hoary old wanderer" made his circumnavigations, even the colonizers' descendants had begun to view both phenomena through a darker lens.

In August 1947, more than three centuries after the East India Company established its first trading posts on the Indian subcontinent, and six decades since the beginning of the Indian independence movement, Great Britain retreated from what had long been the crown jewel of its world-spanning empire. Ten years later, the Gold Coast, soon to be known as Ghana, declared its own independence from Britain, becoming the first sub-Saharan African country to do so and kicking off an era of rapid decolonization that led to continent-wide self-governance in less than a decade.

Along with the post–World War II dismantling of empires came a thorough reconsideration—and broad condemnation—of the colonial project itself. In 1960, the United Nations passed a resolution declaring a fundamental right to self-determination and affirming "the equal rights of men and women and of nations large and small." Among intellectuals, the new field of postcolonial theory gained traction with the 1961 publication of West Indian philosopher Frantz Fanon's influential book *The Wretched of the Earth*. Seeping into the broader culture, these ideas would influence the way people thought not only about the 20th-century world order, human relations, and power dynamics but also about the interrelations of all living things.

An Explosive World

For much of Western history, people had seen nature as a realm to be dominated, controlled, and made useful. In the second half of the 20th century, however, a new paradigm emerged, one in which humans were both part of a complex system known as "nature" and the perpetrators of various harms against it—crimes motivated, like those of colonialism, by hubris and egoism. This idea deeply informed one of the founding texts of modern invasion biology, Charles Elton's 1958 book *The Ecology of Invasions by Animals and Plants*.

Born into an upper-class family of writers and literary scholars in Manchester, England, in 1900, Elton was an introverted and gentle boy, subject to relentless bullying that he later wrote "explains why I have remained uneasy in crowds." As an adult, he remained far more comfortable outdoors; while he would become well-known for introducing several major ecological concepts, the reserved Englishman was never more in his element than when conducting field research far from civilization.

Toward the end of his studies at Oxford, Elton joined an expedition to the Arctic. The trip's leaders included a pair of esteemed British academics, Alexander Carr-Saunders, whose bugbear was human overpopulation, and the evolutionary biologist Julian Huxley, whose grandfather Thomas had been a friend and champion of Darwin. Both men would soon be drawn into the eugenics movement, a social reform effort that aimed to use theories of heredity and selection to shape human populations. It was an early, misguided application of the idea that science could be a force for progressive change. (Julian's brother Aldous would explore the logical extremes of this notion in his 1932 dystopian novel *Brave New World*.)

This first of several Arctic expeditions he would undertake "had a profound influence upon my ideas in ecology," Elton later wrote. The team braved harsh conditions and close calls—Elton once fell through thin ice, drenching himself neck-deep in freezing water—but the young scientist came away with an intimate understanding of ecological fundamentals, which informed his later development of ideas such as food chains, ecological niches, and population cycles. He was particularly fascinated by dramatic population fluctuations in lemmings, which cycle through booms and busts that in turn influence the numbers of arctic foxes, snowy owls, and other predators. The rodents are famously driven to migrate when their numbers exceed their habitat's carrying capacity, a phenomenon Elton witnessed firsthand during a 1930 expedition to Lapland: lying on a riverbank, he watched them "swim across one by one in the faint darkness of the Northern summer."

Along with piquing his intellectual curiosity, Elton's Arctic explorations developed in him a fierce love for wild places and an urge to protect them in the face of global change. This position was personal and political rather than scientific, but it also influenced his theories of invasion biology. Indeed, *The Ecology of Invasions* was fully steeped in conservationist values long before the welfare of the environment had become an issue of broad concern.

"Nowadays we live in a very explosive world," Elton wrote at the very beginning of chapter one, "and while we may not know where or when the next outburst will be, we might hope to find ways of stopping it, or at least damping down its force. It is not just nuclear bombs and wars that threaten us . . . there are other sorts of explosions, and this book is about ecological explosions."

The volume is slim and accessible, punctuated with literary allusions ranging from Edward Gibbon and Walt Whitman to Arthur Conan Doyle's antihero Professor Challenger and the science fiction of H.G. Wells. Elton catalogs "explosions" of chestnut blight, cordgrass, nutria, and sea lamprey, "a bombardment of every country by foreign species, brought accidently or on purpose, by vessel and by air, and also overland from places that used to be isolated." *The Ecology of Invasions* was both a scientific treatise and a lamentation of the loss of "a more pleasant and balanced ecological world."

Writing in the late 1950s, Elton was a little ahead of his time, especially as he argued for the importance of biological variety for its own sake, and for leaving room for wild plants and animals within managed landscapes. But just four years later, Rachel Carson published *Silent Spring*, a foundational text of an environmental movement that would soon become a powerful global force.

Carson, a marine biologist who made a midlife career change to become a popular nature writer, was concerned about the overuse of synthetic pesticides, especially DDT. Although DDT had proven a valuable weapon against malarial mosquitoes, it and related chemicals were also devastating ecosystems, and new evidence showed harm to human health as well. Carson's investigative journalism led not only to the banning of DDT in the United States, but also to the creation of the US Environmental Protection Agency (EPA) in 1970. And she herself credited Elton with influencing much of her own thinking, after his book was brought to her attention by the biologist E.O. Wilson (as noted in Linda Lear's biography of Carson, *Witness for Nature*).

In *Silent Spring*, Carson quoted Elton's *Ecology of Invasions*, noting that healthy ecosystems depend on what the British ecologist had called "the conservation of variety" and connecting invasive species with the overuse of chemicals used by farmers to combat insects and weeds. Such poisons, she argued, would be less necessary if the focus shifted to preventing nonnative agricultural pests from colonizing new territories in the first place. Carson also borrowed Elton's memorable description of these chemicals as a "rain of death," for both her seminal book and for a 1959 letter to *The Washington Post*.

The alarm sounded by early environmentalists like Elton and Carson found full expression nearly a decade after *Silent Spring*, when Senator Gaylord Nelson, a Wisconsin Democrat who had sponsored the legislation that banned DDT, organized the first Earth Day. Modeled after teach-ins against the War in Vietnam, the nationwide protest took place on April 22, 1970, as some twenty million people gathered to call for environmental protection in city centers and college towns from coast to coast. The energy of Earth Day, which carried over into a worldwide annual event, would fuel a series of US environmental protection laws in the early 1970s.

It would also inspire a new generation of scientists.

"After Earth Day, there was more realization and acknowledgment of the importance of the environment," says Mark Davis, a professor emeritus of ecology at Macalester College who has been involved in invasion biology as both a practitioner and a gadfly. "I don't think it's surprising that conservation biology, restoration biology, and invasion biology all kind of emerged at the same time."

Despite Elton's efforts and Carson's attention, invasion biology had remained a niche science for more than a decade after *The Ecology of Invasions* was published. But beginning in the 1970s—and picking up steam in the 1980s—a "mission-driven," or "values-based," approach to scientific research began to take hold among many biologists and ecologists, including those with an interest in invasive species. Unlike Darwin's generation, these scientist-activists no longer saw introduced plants and animals as interesting demonstrations of natural selection at work; now, they were "*Immigrant Killers*," as the title of one 1984 book put it, a problem to be solved rather than simply a phenomenon to be understood.

Founded in 1968, the Elsevier journal *Biological Conservation* published regular studies of invasive species, often appending warnings to policy-makers. In the Pacific, a "plague" of crown-of-thorns starfish had reduced "once-flourishing reef communities to desolate areas of dead coral skeletons," according to a 1970 paper. A 1976 rundown of "exotic" amphibians and reptiles spotted in California cautioned that "raising of pet crocodilians should be emphatically discouraged." In a 1989 paper about a tropical American shrub, Australian researchers predicted "massive loss of animal and plant species . . . if the spread of this aggressive weed is not halted."

The 1980s and '90s saw a continued "flood of publications," as Davis wrote in a 2006 history of invasion biology. There were special symposia, paper sessions at major conferences, and eventually a dedicated journal, *Biological Invasions*, published by Springer beginning in 1999. Scientists and science journalists wrote books aimed at academics and the public, with tones ranging from detached to dismayed, sometimes verging on apocalyptic.

"*Homo sapiens* is, of course, the ultimate invader," wrote Stanford ecologist Paul R. Ehrlich in a chapter penned for the 1986 book *Ecology of Biological Invasions of North America and Hawaii*, which brought together research by some of the biggest names in the field. "Like other 'weeds,' it tends to have a very broad diet. It is also able to use cultural evolution to adapt to very diverse habitats, an ability that complements the loss of estrus which permits it to breed all year round." (Though an esteemed researcher, Ehrlich has remained better-known for his 1968 book *The Population Bomb*, the bestseller that incorrectly predicted humans' mass starvation due to overpopulation and called for coercive methods of birth control—a vision that recalled that of Charles Elton's eugenicist mentors.)

As the field of invasion biology grew, and particularly as its mission-driven faction gained power, so rose the profile of Elton's now-classic work. And scientists, along with the land managers and conservation workers who put their findings to use, were increasingly adopting his rhetoric of war.

"They created a field for themselves, and an identity," Davis says, reached by phone at his home in Minnesota. "And it's enabled lots of people to get funds for research. If you emphasize how dangerous some things might be, and how important it is to learn more, to prevent these horrible things from happening—well, it's more likely to get funded."

The rise of invasion biology, along with its sister fields, conservation biology and restoration ecology, came at a time when environmental awareness was also increasing among the general public. By the 1990s, the scope of the movement's concern had grown beyond pollution and hazardous chemicals to planet-wide change. And as scientific consensus that human activity was causing global warming raised the stakes of ecology and environmental activism, the issue of invasive species also drew the attention of the public, the media, and policy-makers.

There are direct links between invasive species and climate change. Habitats are in flux as seas rise, coasts flood, deserts grow, and permafrost thaws, and, just as in the Richmondian Invasion and the Great American Biotic Interchange, opportunistic species stand poised to take advantage—something scientists have known since Darwin's walk through the Scotch firs. But the connection is also categorical: like carbon emissions, invasive species in the modern era are the fault of humans—especially past generations who, when compared to our educated, environmentally sensitive selves, can appear almost absurdly naïve.

3

AMERICA'S MOST WANTED

In the decades after *Silent Spring* and the first Earth Day, scientists, activists, and popular sentiment worked together to move the US government to adopt new environmental protection policies. This included a dramatic shift in the official American posture toward foreign plants and animals, from one of curiosity—if not rapaciousness—to militant hostility. It's a change that reflects the views of Westerners more broadly, to such an extent that looking back on our recent history now tends to provoke surprise and disbelief.

Today's Americans, along with their counterparts across the industrialized world, have been trained by official messaging to avoid transporting non-native plants like garlic mustard into new environments. At the behest of conservation organizations and government regulators, we wipe our boots at the trailhead in case seeds are hitching a ride in the treads, we scrub aquatic weeds off our boats before launching them into a new waterway, and we surrender foreign fruits and vegetables to government inspectors at the airport.

Not long ago, however, US government scientists and their European colleagues went to great lengths to seek out and bring home new species, in hopes of importing hardier wheat, prettier flowers, or more delicious citrus for farms, gardens, and orchards. Their efforts, a sort of photonegative of today's project to keep foreign organisms out, now amount to little more than a set of interesting footnotes in culinary and horticultural history. But in many cases, they had lasting consequences.

The Plant Hunters

Fanning out from Finland to Fiji, the botanical explorers of the late 19th and early 20th centuries were not only scientists but also adventurers, pursuing the hidden treasures of exotic lands like Indiana Jones—with or without the consent of the locals. Their exploits inspired breathless magazine articles and swaggering personal narratives.

"These Columbuses of the plant world know the remote places of the sphere as few real explorers can," a *Popular Science* reporter enthused in 1922:

> They travel through tropical jungles where boa constrictors, tigers, lions and leopards lie in wait for human prey. They travel on foot through swampy terrain, where the chances of escaping the jungle fevers are 99 to one against the white man. Their only goal, through all this suffering and privation, is a mere plant which, if introduced into the agricultural life of America

may add, it is hoped, to the farm wealth and food resources of the country.

Of course, for many glory was also a motivating factor, and despite their otherwise wonkish botanical interests, plant hunters were loathe to turn down any opportunity to play up the danger of their expeditions to the "remote places of the sphere."

"Few realize the great hardships and dangers which have to be faced in order to secure new plants for cultivation," wrote the Scottish botanist George Forrest, who collected some thirty-one thousand specimens from southern China between 1904 and 1932, at least once disguising himself in local garb to avoid being murdered by Tibetan guerillas. "In the warmer regions there is danger from miasma, fever, animals and snakes. Not infrequently too, the collector has to seek his specimens among savage or semi-civilized peoples, who, in most instances, strongly resent his intrusion into their midst . . ."

Stateside, plant hunting was a well-funded government project. When Abraham Lincoln established the US Department of Agriculture in 1862, he charged it with seeking "new and valuable seeds and plants" that could prove useful as settlers moved westward. Three decades later, the Census Bureau declared the frontier officially closed: Agriculture now stretched from coast to coast, and American farmers were clamoring for new crops to adapt to the host of environments, from high plains to semi-tropical, in which they found themselves. To meet this demand, in 1897 the USDA established the Section (later "Office") of Seed and Plant Introduction, with a twenty-nine-year-old botanist named David Fairchild at the helm. Over the next seventy-five years, USDA explorers—popularly known as "Uncle Sam's Plant Hunters"—would import more than four hundred thousand cultivars into the United States, a global harvest of unprecedented proportion.

Early on, Fairchild realized the USDA needed a place to experiment with the trove of seeds and grafts it was hauling home. He had spent much of the previous decade circumnavigating the globe by steamship in the company of Barbour Lathrop, a wealthy American with a hobbyist's interest in exotic botany. During these travels, Fairchild had developed a special interest in tropical plants, and he knew southern Florida was the only place in the contiguous United States where many of the species he found most fascinating could be bred outside a greenhouse.

USDA higher-ups expressed less enthusiasm about the Sunshine State than did Fairchild and his staff. Florida in 1898 was something of an agricultural backwater, wholly unsuited to the staple grains that, along with beef, made up the bulk of the American diet. The state's citrus industry, worth nearly $1 billion today, was just getting started, and despite Fairchild's own adventurous palate, there was little sign that Americans would embrace new types of fruit. Miami, today the heart of a metropolis that's home to six million people, then consisted of little more than a few hotels clustered around a railroad terminus. But when local landowner Mary Brickell offered six acres of wilderness for a study garden along the Miami River, Fairchild took a train south, then bicycled along a sandy road through the hardwood hammock to check it out.

"After the East Indies, the Florida vegetation seemed a little disappointing in its lack of height, but it was dense enough to be called a jungle, and certainly was tropical in character," he wrote in his 1938 memoir, *The World Was My Garden*. "To us young fellows, this six acres seemed quite marvelous. It was the first land available to us where we could plant anything we wanted."

And plant they did. During Fairchild's tenure, the Office of Plant and Seed Introduction notched many notable triumphs, including clementine oranges and Meyer lemons, multiple types of avocado (then known as "alligator pear"), hardy durum wheat, and forty-two new varieties of soybean. Some imports, although given a fair shake, failed to catch on immediately. Kale and quinoa, among the hippest foods of the early 21st century, were initially met with suspicion by American consumers, who considered raw celery an exotic dish. In other cases, despite the USDA's best efforts, growing conditions in the United States were simply unsuitable. Fairchild long lamented the failure of the mangosteen, an "indescribably delicious" fruit "with a sprightliness of flavor" that he tasted during an 1890s stopover in the East Indies.

"Of course, I immediately wanted to see this fruit on the American market, but there were many difficulties to be overcome," he wrote. For one thing, "Java in those days was almost as distant as the moon."

Other introductions later proved to be regrettable mistakes. According to the USDA's contemporaneous inventory, in 1903 the department acquired seeds of the Chinese tallow tree (*Triadica sebifera*). It was not the species' first foray into North America; posted in London as the colonial representative to the crown in 1772, Benjamin Franklin himself had picked up a few seeds at an expo and mailed them to a friend in Georgia. The tallow tree appeared to be "a most useful plant," the founding father wrote, noting that its waxy seed coating might serve as a substitute for beef fat in soap and candles.

Franklin's specimens flopped, but the USDA would have more success—too much, really. Hoping to spark the development of a new fuel oil industry, the department had tallow seedlings planted in Texas. By the end of the 20th century, the versatile hardwood had spread through the Gulf states, transforming native prairies and marshland alike into groves of tallow trees.

It was not the only USDA-introduced species to get out of hand. In 1914, researchers imported seeds of the Australian cajeput tree, *Melaleuca quinquenervia*, and distributed the seedlings in Florida. Fairchild later expressed contrition for this well-intended venture: "Who could have imagined that from a row of them . . . there should grow an actual forest of Cajeput trees which would destroy even the citrus grove that had been planted close to them?" he wrote in 1938 remarks to the Florida State Horticultural Society. "Is someone to see the Cajeput tree become one of the character trees of the Everglades, I wonder."

By then, Fairchild's colleagues elsewhere in the federal government were doing all they could to make these speculations come true. In the 1930s, the US Army Corps of Engineers planted groves of melaleuca, which can shoot up six feet in a year, to stabilize levees in the Everglades. At least one pilot broadcast seeds from the air as part of an effort to dry out what was generally seen as a "useless" swamp. One hundred years after the USDA imported it, melaleuca had taken over half a million acres in South Florida.

Fairchild personally helped spread another of America's most notorious weeds, a plant that in some states is now synonymous with invasive species. In his travels, Fairchild had seen Japanese farmers feeding a massy vine they called "kudzu" (*Pueraria montana*) to cattle. Intrigued, the esteemed botanist had his staff plant some at the back of the Miami garden (a patch that proved so vigorous it soon "threatened to drive our neighbors out") while he experimented on his own at the wooded forty-acre estate in Maryland that he and his wife Marian purchased shortly after their 1905 wedding.

"I grew a lot of seedlings and scattered them about rather recklessly," Fairchild wrote in his memoir. "The seedlings all took root with a vengeance, grew over bushes and climbed the pines, smothering them with masses of vegetation which bent them to the ground and became an awful, tangled nuisance." Although the Fairchilds spent more than $200—roughly $3,000 today—on efforts to remove the tangle of kudzu, enough remained when the family sold the place in 1924 that the new owner was able to pasture his cow on it all summer. (Later, the USDA's Soil Conservation Service began using kudzu to control erosion—a scheme that would soon prove ill-advised, as the vine nicknamed "mile-a-minute" spread through the southeastern United States.)

Meanwhile, the Miami Plant Introduction Garden kept growing, too. By 1914, the original site had become too crowded, so the department leased another twenty-five acres in the subdivision of Buena Vista. But both sites were soon hemmed in by the growing city.

"I began to realize that 25 acres would be totally inadequate to contain the trees and the various activities of a Federal Garden which was being fed by the tropical explorers . . . at the rate of several thousand new introductions a year," Fairchild said. "Something would have to be done."

Fairchild, who could be annoyingly persistent where his plants were concerned, eventually persuaded Secretary of War John Weeks to let the USDA use Chapman Field, an abandoned World War I airbase five miles south of the city. The entire garden operation moved there in 1923, and although its most dogged champion retired in 1935, new species continued to arrive through the 1960s at the rate of five hundred a year. All along, the busy facility—which eventually expanded to 125 acres—was separated from the Everglades by little more than a chain-link fence.

Running between one of the country's most sensitive ecosystems and the wild-eyed experimentation of America's globe-trotting plant hunters, that hilariously permeable barrier now belongs firmly to another era. Indeed, in the decades after Fairchild's tenure, the USDA's posture toward non-native plants and animals became increasingly defensive. The department had maintained the authority to inspect agricultural imports for pathogens since Congress passed the Plant Quarantine Act in 1912, but after World War II the prevention of pest introduction took up more and more of the department's attention, as non-native species like Japanese beetles (*Popillia japonica*) and gypsy moths (*Lymantria dispar*, renamed "spongy moths" in 2022 to avoid racist implications) posed a growing threat to US agriculture and forests. Plant introduction was eventually subsumed into the National Plant Germplasm System, the mission of which is to preserve genetic diversity in the face of a host of ecological threats, rather than to seek out novelty in a world full of wonders.

In 1971, just a year after the first Earth Day, the USDA created a new agency tasked with responding to the growing problem of agricultural pests and diseases with origins abroad, to eventually be dubbed the Animal and Plant Health Inspection Service (APHIS). Among other programs, its $2 billion budget covers the Beagle Brigade (those trusty hounds who sniff out foreign food in the baggage of air travelers) and research to combat the spread of introduced species like melaleuca and tallow trees. The department's scientists are still hard at work in South Florida—but now instead of hunting for marvelous new fruit from around the world, they are seeking new ways to eradicate some of the very species their forebears introduced.

Aliens Among Us

Once upon a time, the press had fallen hard for Uncle Sam's Plant Hunters. But in the late 20th century, it pivoted along with the USDA and found an even better story, one that offered a clear, good vs. evil narrative, colorful characters, a simple take-home message for readers and viewers, and—most deliciously—the opportunity to indulge in purple prose.

A 1995 *New York Times* article headed "Attack of the Aliens: Florida Tangles With Invasive Species" was typical of the genre, warning about "a growing army of foreign plants and creatures, some of them highly invasive, from South American citrus to disease-bearing African monkeys and Asian mosquitoes, plus an occasional free-roaming boa constrictor." In what was once David Fairchild's backyard, "unique native flora and fauna" were "being quietly displaced by the newcomers," wrote the reporter, Alan Burdick. (Continuing on what would prove to be a rich beat, Burdick went on the author the National Book Award finalist *Out of Eden: An Odyssey of Ecological Invasion* in 2005.)

Journalists often hooked audiences with appeals to the universal human fear of monsters, before introducing a dramatic twist: the article was not *literally* about the marauding army of a hostile power or a creature from another planet, but a humble-seeming vine or rodent. Indeed, the contrast between the apparent ordinariness of invasive plants and animals and the threat they posed served to heighten the sense that invasive species were not only ecologically inconvenient but also somehow dishonest. ("Imagine if someone broke into the Metropolitan Museum of Art," a biologist told Burdick—talking about weeds, to be clear—"and replaced all those priceless artifacts with fakes.")

It was a rhetorical strategy with perennial appeal. A few years after the *Times* sounded the alarm about Florida, for example, *Washington Post* staff writer Joel Achenbach surveyed the situation in his native state by helicopter, writing with obvious relish about

> a pale-green substance that seems to be crawling all over the tree islands that speckle this portion of the Everglades. The pilot takes the chopper down for a closer look. You can see it, sure enough: lygodium. Old World climbing fern. It has gone

> berserk. It's like the Blob. The islands are caving in at the center, crushed by the dense, matted blanket of vegetation. The willows, the hollies, the cabbage palms—they're being buried alive.

Scientists, land managers, and government officials were often willing partners in such reporting, supplying killer quotes as they worked diligently to make the case for public concern—and public funding. "It will kill everything beneath it. . . . It's bad news. Bad news," one biologist insisted to Achenbach about giant salvina (*Salvinia molesta*), an aquatic plant native to Brazil. "I think metastatic cancer is the strongest analogy," a US Geological Survey plant ecologist added soberly.

National newspapers were joined by local publications and TV stations, for which stories of newly sighted aliens served as a sort of proto-clickbait. One snakehead (*Channa argus*) spotted in a single Maryland pond prompted a weeks-long news cycle, during which everyone from the *Baltimore Sun* to CNN speculated with horror about the dastardly intentions of this "ghastly fish," a carnivorous Asian import that can live outside water for several days. (The drama ended when the pond in question was doused thoroughly with poison, killing snakeheads and native fish alike.)

Snakeheads were not the only tabloid-ready monster that terrified Americans during the 1990s and early 2000s. "Remember how scared we were of killer bees?" a Reddit user mused recently in a forum dedicated to Generation X nostalgia, prompting coeval respondents' comparisons to childhood fears of nuclear war and kidnapping.

So-called Africanized honeybees originated in Brazil in the 1950s, when scientists attempted to crossbreed European bees (*Apis mellifera*) with a more heat-tolerant African subspecies (*Apis mellifera scutellata*) in order to boost their tropical country's honey production. In 1957, two dozen African queens escaped from captivity and established feral hives with European honeybees outside São Paulo. Over the next few decades, their offspring expanded northward, arriving in Texas in 1990 and California in 1995.

During their decades-long march north, the hybrid bees—which tend to sting more aggressively and defend wider territories than European bees—were blamed by officials for killing some one thousand people, earning them the nom de guerre "killer bees." In addition to their deadly reputation, authorities in the United States also worried that Africanized bees could infiltrate domestic apiaries, threatening a $140 million industry.

While a handful of fatal bee sting cases made the news once the bees finally reached the United States, public fears never fully came to fruition. (Detailed statistics are hard to come by, but for context, CDC data shows just under 2,300 Americans died after being stung by *any* type of hornet, wasp, or bee between 1979 and 2020, with annual mortality rates fairly consistent year over year.) Africanized bees are now common across the southern quarter of the country, and it's impossible to tell whether any particular colony is made up of "killer bees" or their more docile cousins—one feral bee colony turns out to look a lot like any other, and scientists have largely given up on routine testing for "Africanized" genes. Meanwhile, lost in the panic over this invasive subspecies was the fact that the bees upon which the US honey industry depends—insects that are also vital for pollinating commercial crops and orchards—are themselves imported, brought from Europe by settlers during the 1600s.

In recent years, honeybees themselves have become the subject of a different concern—this time, a mysterious phenomenon called Colony Collapse Disorder, connected variously to pathogens, pesticides, or other stresses, which causes hives to fail when workers flee. While the syndrome has become less common than at its peak around 2008, it has merged with worries about the global decline of pollinators in general, to the irritation of scientists who point out that honeybees are, in fact, competitors with native bees for nectar and pollen, and can even spread diseases that threaten native species directly.

"[B]eekeeping is an agrarian activity that should not be confused with wildlife conservation," grumbled Cambridge zoologists Jonas Geldmann and Juan P. González-Varo in a 2018 op-ed for the journal *Science*. "There is widespread concern about the global decline in pollinators.... Yet, concern has focused on one species above all: the western honey bee.... This has led to initiatives, masked as conservation, that promote honey bees in cities and even in protected areas far from agriculture."

Geldmann and González-Varo noted that honeybees may have an adverse impact not only on native bees but also on native plant communities, because they interfere with their preferred pollinators, and even certain birds, with which they compete over nesting space. While thirty states have passed laws protecting honeybees, such efforts are more akin to conserving dairy cows or soybeans than saving the spotted owl.

It's a message that has curiously failed to reach the average concerned citizen—my Ann Arbor neighbors, for example. Proud participants in the nouveau tradition of No-Mow May, they persist in posting lawn signs bedecked with cheerful black-and-yellow insects—distinctively, honeybees—and the entreaty "Pardon our weeds. We are feeding the bees."

Which environmental stories catch on with the public, which species are held up as heroes, and which are smeared as villains often comes down to the emotional appeal of the plant or animal in question: the killer bee from darkest Africa vs. the friendly honeybee, the freakish snakehead vs. the all-American largemouth bass. Narratives develop without nuance, especially when there may be little time to spare before urging the public to action. And while inspiring panic about oddball fish and scary bees may be low on the list of the media's crimes, such reporting takes on a more sinister cast when it echoes the language used to discuss another hot topic of the late 20th and early 21st centuries.

Stampedes

In November of 1986, President Ronald Reagan signed the Immigration Reform and Control Act. It was the first major overhaul of immigration law in the United States since 1965, when Congress did away with a quota system based on national origins that had been in place since the 1920s.

A compromise between immigration hardliners and civil rights advocates, the 1986 law placed sanctions on employers who knowingly hired any "alien" not authorized to work in the United States, while at the same time extending amnesty and a path to legal citizenship to some three million people who had been living in the country illegally for at least four years. The law, Reagan declared at the signing ceremony, would "go far to improve the lives of a class of individuals who now must hide in the shadows, without access to many of the benefits of a free and open society."

Despite the president's apparent magnanimity toward undocumented immigrants, for much of the 1980s they had been the subject of concerned op-eds and tense congressional hearings. Politicians railed against illegal border crossings, often by refugees who were fleeing civil wars and unrest in Nicaragua, El Salvador, Honduras, and Guatemala—the so-called Central

American Crisis of the late 1970s. This anxiety flared up all along the political spectrum, spurred by fears about crime, competition for jobs and public resources, and cultural change. All of it was rooted in the idea that native-born Americans didn't stand a chance against a flood of new arrivals.

"Our immigration policy is making us poorer, not richer. . . . It is dividing our wealth and resources," worried then-Governor Richard Lamm, a Democrat from Colorado. In a 1982 report on immigration for the Senate Judiciary Committee, Senator Chuck Grassley, a Republican from Iowa, summed up the position of immigration hawks: "Though most Americans currently residing in the United States owe their presence here to the open door immigration policies of the past[,] the perpetuation of the myth that this country is one without limits can only prove harmful in the long run."

Grassley went on to quote the Pulitzer Prize–winning political journalist Theodore H. White's 1982 bestseller *America in Search of Itself*, which described changes in transportation technology that had sped up immigration, from the weeks endured by migrants from Europe in the 19th century to, in the late 20th century, mere hours by plane from places as far away as China, India, or Africa. "The result has been a stampede, almost an invasion," White wrote. As a result, "the United States has lost one of the cardinal attributes of sovereignty—it no longer controls its own borders."

Even pundits who supported immigration hinted at the possibility that existing Americans might have difficulty competing with these hardworking newcomers. "The single most important quality in immigrants is the willingness to adapt," wrote journalist James Fallows in a 1983 cover story for *The Atlantic* that profiled an extended family from Vietnam who owned multiple homes and businesses a few short years after their arrival in

California as penniless refugees. "The immigrants themselves, compared with native-born people of the same race and with the same amount of schooling, start out at a big earnings disadvantage. But, in a matter of years, even the first generation catches up with and then passes the native-born."

Reading such stories about human immigration, anyone with invasion biology on the brain can't help but recognize similarities between the way Americans talk and think about both subjects: the vigorous stranger pitted against the vulnerable native, the fear that long-established systems could be upended by unauthorized additions. Such parallels certainly do not escape Banu Subramaniam, who traced these connections in her 2014 book about nature and culture, *Ghost Stories for Darwin*: "[T]hese are all symptoms of the same anxieties, the same problem—our cultural anxieties of a purportedly fast-changing world, anxieties that we and 'our kind' will be left behind."

A successful immigrant herself, Subramaniam left India for the first time in 1986 to pursue a graduate degree in genetics at Duke University in North Carolina, "with visions of donning that revered white lab coat amid a sea of white-skinned, white lab-coated male scientists studying the proverbial white male rat." But as she investigated color variation in the common morning glory, she began to feel increasingly alienated from the detached, sometimes self-exalting culture of scientific research—a culture that, she realized, failed to recognize itself as a culture at all.

Subramaniam dabbled in women's studies as she completed her PhD at Duke, and for a time she kept one foot in biology and the other in the humanities, researching floral polymorphism and the sociology of women in science, Hindu nationalism and the relationship between soil-dwelling organisms and invasive plants in California. But she soon began to see divisions between the world of science and the world of culture, and between nature and human beings, as artificial.

"I think there is something about Western science and the Western imagination that has this hierarchy of beings, where the human and non-human worlds are separated," Subramaniam says, reached at her office in Wellesley, Massachusetts, where she has been the chair of women's and gender studies at Wellesley College since 2023.

It's that binary, she believes, that allows liberals to see the right-wing anti-immigration position as morally repugnant, while opposition to invasive species remains an uncontroversial tenet of standard-issue progressivism. And it's what lets people on the right, who consider border security paramount, dismiss environmental regulations that might tamp down biological invasions as a threat to freedom. Meanwhile, both sides continue to tell similar stories about outsiders who are—as Subramaniam writes in *Ghost Stories*—"everywhere, taking over everything" and "silently growing in strength and numbers."

Is there something universal in human nature, some fundamental drive for purity, from which these two apparently opposite positions ultimately spring?

"I think the roots of the two are slightly different," Subramaniam says, cautiously. "Nativism toward plants comes from environmentalism." Still, within the movement there has always been a yearning for "pristine wilderness," she notes, a desire that, at the deepest level, has much in common with nationalist sentiments about an uncorrupted culture. "They're entangled way back in the past," she says, "but I do think environmentalists would be shocked to think that there might be connections—'Don't tell me I'm a white nationalist!'"

The crunchy granola environmentalist and the flag-waving immigration hawk may struggle to believe they have anything in common—and would be appalled to find they do. But in their different ways, both are motivated by a fear of global change in a technology-driven era when everything seems to be sped up and happening all at once. It's an anxiety that seemed to reach new heights in the United States at the turn of the millennium and that would continue to influence government policy on both issues for decades to come.

Back on the Hill

In the fall of 1996, just ten years after the last attempt at comprehensive immigration reform, President Bill Clinton signed another sweeping law. The Illegal Immigration Reform and Immigrant Responsibility Act rolled up the welcome mat that had been laid out in 1986, an amnesty offer that had since helped some three million immigrants gain legal status. The new rules would make it easier to send the undocumented back to their countries of origin, add additional penalties for those who committed crimes while in the United States, and increase enforcement along the southern border. By 2000, nearly two hundred thousand people were being deported annually, a sixfold increase over a decade.

The 1996 immigration measures were controversial, but another piece of legislation, passed less than a month later and targeting a different set of newcomers, made fewer waves. The National Invasive Species Act of 1996 called for new regulations and research and enlisted the US Coast Guard to prevent the spread of "nuisance species" in the Great Lakes, the Chesapeake Bay, and other American waters, with a specific focus on zebra mussels, which were originally introduced from Eurasia in the ballast waters of cargo ships.

A law targeting aquatic invasives, however, soon proved insufficient to assuage the concerns of scientists and the public about a rising tide of invasive species arriving on American shores. In a 1997 letter to Vice President Al Gore, some five hundred scientists, conservationists, land managers, and agriculture officials demanded further action to address "a rapidly spreading invasion of exotic plants and animals" that was "destroying our nation's biological diversity" and "costing the U.S. economy hundreds of millions of dollars annually." That letter was followed by another shortly afterward from Senator Bob Graham of Florida, who urged the administration to do something about "the growing environmental threat posed by alien (non-indigenous) invasive species."

Gore, long a passionate environmentalist, was receptive to the message—in fact, as he told Graham, the White House had already organized a group of its scientific advisors, known as the Biodiversity and Ecosystems Panel, to look into this and other threats to America's "natural capital."

The document produced by that panel in 1998 was titled "Teaming with Life," but the cute headline belied the report's dire picture of destruction and loss that imperiled both biodiversity and civilization. Citing threats that included "killer bees, zebra mussels, fire ants, and the Mediterranean fruit fly," the report warned that "[w]ithout far-reaching changes in the quality of our stewardship of our natural assets, problems of this sort will escalate in both number and intensity as human populations and consumption of goods and services increase."

The report was well-received by the Clinton administration and by the many interests that influenced it—indeed, mounting a powerful counteroffensive against invasive species seemed to be something that everyone could agree on. The Mediterranean fruit fly, after all, was not in a position to hire a public relations team to defend it, and powerful agricultural and industrial interests were all in when it came to fighting crop pests and "noxious weeds." That seemed especially true when the battle required large orders of pesticides (as the journalist Andrew Cockburn pointed out in a 2015 *Harper's* story, the biodiversity panel's chair, botanist Peter Raven, spent much of his career working closely with the American agrochemical manufacturer Monsanto).

In the background, however, loomed another environmental threat, one that the 1998 report only touched on and which would prove far more resistant to consensus.

Al Gore had learned about global climate change long before the five hundred scientists brought the invasive species issue to his desk. As an undergraduate at Harvard in the late 1960s, the future vice president took a course with oceanographer Roger Revelle, who had demonstrated a decade before that carbon dioxide was rising in the atmosphere. After his election to Congress in 1976, Gore held the first hearings on climate change, inaugurating decades of activism that would culminate in his being awarded the Nobel Peace Prize in 2007. But Gore's strong feelings on the matter were never sufficient to move his government, and in 1997 the United States was among just a few nations that declined to ratify the Kyoto Protocol, the first international agreement designed to fight climate change by requiring industrialized countries to reduce their emissions of carbon dioxide and other greenhouse gases.

Climate change is a multifaceted problem, without a single obvious villain. Taking on non-native plants and animals, on the other hand, will always be an easier lift, even if wiping them out is at least as unrealistic as holding down global temperatures. It's more politically palatable, after all, to declare war on garlic mustard—a weed that was never likely to inspire much loyalty—than to target carbon-intensive industries, with their armies of lawyers and lobbyists, or indeed the very carbon-reliant lifestyles to which modern people have become accustomed.

In any case, no outcry was heard on February 3, 1999, when President Bill Clinton—at a ceremony with some of the scientists who had written to Gore—signed Executive Order 13112, calling for a coordinated federal response to unwanted foreign plants and animals. The order requested federal appropriations and created the National Invasive Species Council, which would bring together representatives from thirteen federal departments and agencies to create a comprehensive national strategy for prevention, mitigation, and, wherever possible, wholesale eradication—by any means necessary.

"This is a unified, all-out battle against unwanted plant and animal pests that threaten to wreak major economic and environmental havoc," said then-Secretary of Agriculture Dan Glickman, one of the council's co-chairs.

David Fairchild might hardly have believed it, but a century after he laid out his experimental garden, the United States was at war.

4

A VIEW FROM THE TRENCHES

Jim Hurley and Susan Roth will tell you that they knew what they were getting into.

In 2013, when the couple bought 156 acres of woodland and meadow in the foothills of Virginia's Blue Ridge Mountains, their plan was to gain a beachhead in the war on invasive plants to show what could be accomplished with handsaws, gardening gloves, a little herbicide, and a whole lot of determination. A decade later, they enjoy a great sense of satisfaction as they stroll among native spring ephemerals—toothwort, rue anemone, hepatica, mountain bellwort, and wild geranium—and carpets of native fern beneath a closed canopy of hickory and oak. Everything they can see belongs right where it is.

Maintaining that harmony, however, has required constant vigilance.

The first enemy the couple confronted on their land was Japanese stiltgrass (*Microstegium vimineum*) spreading along the top of the steeply sloped property. Runoff from a gravel road had carried its seeds hundreds of yards downhill, and the sprawling annual formed dense stands in the woods. So Hurley and Roth hiked up with 3-gallon, backpack-mounted sprayers and doused the weeds in herbicide. Then they tangled with thorny thickets of multiflora rose (*Rosa multiflora*). Once that was under control, they began noticing Oriental bittersweet (*Celastrus orbiculatus*), a woody vine that can climb sixty feet and spread through the canopy.

"Bittersweet is horrific," Hurley says, shaking his head. "And now there's a terrifically horrific new one, *Youngia japonica*, Asiatic false hawksbeard. In the months of March and April, I have to spend probably ten hours a week trawling, just on the three or four acres that I know that it's on. And I have to go back again and again and again and again."

To call such an effort Sisyphean feels a little too on the nose, but the allusion is impossible to resist, especially as new invasive species continue to spread downhill thanks to less scrupulous neighbors upslope (among them, the state of Virginia, which maintains that troublesome gravel road). Hurley, a retired organizational consultant, estimates that he personally spends between five hundred and seven hundred hours a year combating invasive plants, on top of the bigger projects he has contracted out—"and that's well north of $100,000."

It helps that he and Roth are not alone. They're part of a platoon tackling invasive plants at all levels, from their own backyards all the way to Richmond, where they've recently made some legislative headway. The group is called Blue Ridge PRISM

("Partnership for Regional Invasive Species Management"), and Hurley, who is on the nonprofit's board, jokes that it has a two-part mission.

"One is to reduce the impact of invasive species in the landscape," he says. "Our secondary mission is to ruin your day. That is, to open your eyes to all the dreck that's around you. And once you see it and you know what you're looking at, you can't not see it."

What's really frustrating is when people refuse to see it the way PRISM does.

Twenty-five miles south of Hurley and Roth's place, a pretty drive along winding mountain roads, sits the small city of Crozet (pronounced "Croz-ay," it's named for a French-born civil engineer who designed railroad tunnels through the Blue Ridge Mountains). A bedroom community for Charlottesville, which is home to the University of Virginia, Crozet boasts what has become a rarity in the United States: its own local newspaper, the monthly *Crozet Gazette*, with a circulation of five thousand.

With a staff of four and a rotating cast of freelancers, the *Gazette* covers city and county government, high school sports and human interest, and it publishes columns by locals with expertise on everything from Virginia history to cybersecurity. For more than a decade, one of the most regular was "The Blue Ridge Naturalist," which offered tips for readers hoping to make their yards more hospitable to wildlife.

Its author, Marlene Condon, typified the small-town environmentalist: she was concerned about climate change and the fate of pollinators, and she bemoaned what she saw as overdevelopment, along with other threats to the natural world she loved. In between warnings about pesticides and lyrical observations about the birds, mammals, and amphibians who visited her garden, however, Condon also staked out more renegade positions. When a northern copperhead made its den under her carport, she wrote about learning to live with the venomous snake and urged readers to overcome their own herpetophobia. She landscaped with brush piles and suggested leaving table scraps outside to feed "our natural sanitation workers," opossums and raccoons.

From time to time, Condon also expressed skepticism about the dangers posed by invasive species. Far from being a universal scourge, she argued that such quick-growing plants served a useful purpose as food and shelter for native animals, especially in disturbed or degraded areas where more delicate natives struggled to take root. Non-native species like the Eurasian shrub autumn olive (*Elaeagnus umbellata*) grew unmolested in her own garden, where she wrote about watching butterflies sip their nectar and birds feast on their fruit.

But a column that ran in February 2019 apparently took this *laissez faire* approach a little too far.

The news peg that month was California's plan to remove Australian eucalyptus, a non-native tree that monarch butterflies—a beloved and vulnerable species—had adopted as a roosting site. Beneath the headline, "Ecologists Recognizing Value of Alien Plants," Condon raised a figurative eyebrow at this "deliberate destruction of habitat." The move was based on blind ideology, she alleged. "Native-plant folks out west have managed, as they have here in the East, to convince environmental organizations and government entities at every level that it is a moral imperative to remove plants deemed invasive."

PRISM members felt provoked.

"It got my blood boiling," says William Hamersky, a PRISM member who lives just south of Crozet. "There are stacks of the *Gazette* in most stores. There are some Crozettians—Crozeeshans?—that I'm sure read it cover to cover. So it was just miseducating the public."

Hamersky relocated here in 2015 from the San Francisco Bay Area, where among other jobs he played a French knight at Renaissance festivals, participating in jousting matches and swordsmanship demonstrations. Though he still sports a rakish gray ponytail and Van Dyke beard, "Sir Guillaume" has laid aside his lance for a new weapon of choice, wielding a machete against invasive species like porcelain berry (*Ampelopsis brevipedunculata*).

In the days after Condon's column ran, feverish emails bounced between PRISM members, and soon Hamersky had teamed up with Roth—who is a professional writer with a background in horticulture—and a University of Virginia biologist, Manuel T. Lerdau, to pen a scorched-earth letter to the editor, which appeared in full in the next month's *Crozet Gazette*. Over 1,800 words, the trio laid out their case against both Condon herself ("She engages in blanket and baseless accusations and name-calling . . .") and invasive plants like garlic mustard and Oriental bittersweet, invaders that they argued "are not members of our local ecosystems and plant communities." They quoted the USDA and cited scholarly research. They asked readers to "consider Condon's diatribes with skepticism."

Condon got a chance to respond in the next month's issue, but the PRISM letter proved a death blow to the Blue Ridge Naturalist; that spring, after more than a decade, the *Gazette*'s editor abruptly terminated the column.

The Heretic

The geographical origin of species may be an especially touchy issue in Albemarle County, which is best known to Americans outside the region as the home turf of the Founding Fathers Thomas Jefferson and James Madison. People here place a special value on historical landscapes, and the warm, wet climate seems, anecdotally at least, to welcome Eurasian imports ("Virginia is so green, *everything* grows here," Hamersky laments). But the canceling of the Blue Ridge Naturalist was a hyper-local echo of a broader, long-running controversy that has divided not just gardeners and landowners but also international academics into seemingly intractable factions.

Almost since its inception, invasion biology has weathered criticism from journalists, social scientists, philosophers, and humanities scholars like Banu Subramaniam, who are drawn to the ethical and ontological questions dredged up by efforts to divide organisms into native and non-native and to exclude or eradicate the latter. But internal criticism has also dogged the field, perhaps most notably a 2011 commentary in the journal *Nature*, "Don't judge species on their origins." It put forth an argument for toning down rhetoric against non-native species:

> It is time for scientists, land managers and policy-makers to ditch this preoccupation with the native-alien dichotomy and embrace more dynamic and pragmatic approaches to the conservation and management of species. . . . Classifying biota according to their adherence to cultural standards of belonging, citizenship, fair play and morality does not advance our understanding of ecology. Over the past few decades, this perspective has led many conservation and restoration efforts down paths that make little ecological or economic sense.

The notion that native species are inherently good, useful, or somehow supposed to be where they are has no evidence behind it, the authors argued, and in fact relationships between native and non-native species are often far more complex than the "invasion" model assumes. The native mountain pine beetle (*Dendroctonus ponderosae*), the authors noted, is responsible for killing more North American trees than any other insect pest, despite reams more news coverage generated by foreign imports like the emerald ash borer (*Agrilus planipennis*). Meanwhile, they wrote, the US government had allocated $80 million over five years to eradicate various species of invasive tamarisk shrub (genus *Tamarix*)—which turned out to provide much-needed habitat for the endangered southwestern willow flycatcher (*Empidonax traillii extimus*). While some non-native species cause disruption, the authors argued for basing decisions about management on their function within an ecosystem rather than whether they existed there before humans became a factor.

Published by one of the world's top scientific journals and co-authored by nineteen scientists—including *éminences grises* with decades of experience on the front lines of invasion biology—the commentary was well-credentialed. But perhaps unsurprisingly, given the high-profile challenge it presented to an entire scientific subfield, the reaction against it was swift.

A month later, the issue dominated *Nature*'s "Correspondence" section, as other academics wrote in to object to what they saw as an "unfair" characterization of invasion biology and to urge continued vigilance against non-native species, both those that were and those that might become invasive. One letter, published under the sub-headline "141 scientists object," was signed by a who's who of invasion biologists and ecologists at universities, nonprofits, and government agencies around the world—including *Population Bomb* author Paul Ehrlich, the Berkeley ecologist who had called humans "the ultimate invader."

"It was a very short paper, two pages," says Mark Davis, the original essay's first author, with a sigh. "It wasn't even a full two pages, because there were some pictures. But I don't think most people who criticized it read the damn thing."

Davis also retains hard feelings over the title chosen by *Nature*'s editor (Don't judge species on their origins). It's an oversimplification of what he and his colleagues were trying to say, he insists. "We were just calling for a change in emphasis. We weren't saying origin doesn't matter at all. We just wanted more focus on what's actually happening, versus where the species is from. But it's been misrepresented even by people that support that perspective, which is really discouraging and kind of disappointing."

Part of the problem may have been that Davis's reputation preceded him.

The since-retired Macalester College ecologist, whose 2006 history of invasion biology traced the discipline's development into what he called a "values-based" field, had gotten his start as an entomologist and plant ecologist, before making his name in invasion biology with a 2000 paper that proposed a theory of "invasibility," a measure of how susceptible a plant community is to colonization by non-native species. Davis, along with colleagues at the University of Sheffield in the United Kingdom, showed that a plant community will become more susceptible to invasion by newcomers when more resources—sunlight, water, soil nutrients—become available, whether because resource use by preexisting flora declines or because the supply of resources increases faster than local plants can use them. Broad theories about biological invasion have proven more elusive than the field's pioneers originally hoped, and Davis's paper was elegant and influential. It has since been cited more than four thousand times.

Even as Davis was finding success in invasion biology, however, something about the whole concept had been bugging him.

In 1994, an article by the journalist Michael Pollan (today best known for his pithy dietary prescription, "Eat food. Not too much. Mostly plants.") appeared in the *New York Times Magazine* under the headline "Against Nativism." At the time, natural gardening was a growing movement, its practitioners advocating the creation of landscapes that were as close to untouched nature as possible, composed solely of native plants that offered food and refuge to native animals. Pollan, who had recently written a book-length meditation on gardening, *Second Nature*, questioned the movement's "intolerance toward foreign species." He connected it with historical antipathy toward human foreigners and noted the futility of undoing the "mongrel ecology" humans have created over centuries of our own migrations. "Turning back the ecological clock to 1492 is a fool's errand," Pollan wrote.

"I read that article," Davis says. "As an ecologist, I was very pro-native species and against non-native species. It angered me that he would make that link with nativism. But in the succeeding days or weeks I just couldn't get that out of my head. It was kind of like a pebble in my mental shoe—it just kept bothering me. I felt I couldn't really counter his arguments. Eventually, I realized, 'Damn it, he's right.' And once I started that shift myself, I started to pay attention to the language, to what people were saying at conservation conferences, and it came to the point where I couldn't just stand by as this really bad science was being promoted. I just felt that I needed to critique it. And so that's what I did."

Davis began with gentle suggestions for defining the field's terms more precisely, which he argued would help invasion biology become a more "quantitative, analytic, and predictive science." In 2000, Davis and the British ecologist Ken Thompson, who was also a coauthor on Davis's "invasibility" paper, proposed an eight-part typology to help differentiate "invasive" species—which may require management—from those that were merely "colonizers," arguing that the former label ought to require not just presence in a new environment but a demonstrated impact on other species in its adopted range.

Some of Davis's colleagues pushed back. The mere fact that a non-native organism was spreading in a novel environment, they said, ought to be enough for biologists to use the term "invasive," with all its implications. After all, by the time an impact could be proven, it might be too late to do anything about it.

In response, Davis and Thompson doubled down, asserting that invasion biology was a "pseudo-discipline" that lacked justification for its separate identity as a subfield. Indeed, they insisted in a letter published by the *Bulletin of the Ecological Society of America*, invasion biology had been able to differentiate itself "largely as a result of effective promotion and the use of discipline-specific jargon," which created the erroneous impression "that the introduction (and occasional spectacular success) of new species represents an ecologically unique phenomenon."

"We fear that, despite original good intentions, the emergence of invasion ecology as a distinct subdiscipline has hindered more than helped," Davis and Thompson wrote, and suggested that science would be better advanced by freeing investigators from the separate identity of "invasion biologist" that had developed since the 1980s.

Continuing to needle the establishment, in 2003 Davis challenged one of the most-repeated axioms of invasion biology, that invasive species represent the second-greatest extinction threat after habitat loss. While introduced predators or pathogens have indeed wiped out naïve species in isolated habitats such as islands and lakes, Davis, writing in the journal *BioScience*, reported finding no cases in which non-native species had caused extinction through competition, despite the common warning that invasive species will "crowd out" natives.

(Two decades later, there is still no solid evidence that competition *alone* has driven any species to extinction.) With this, Davis cheekily dismissed the worry of invasion biologists that the future would bring "a world composed of zebra mussels, kudzu, and starlings."

Over time, Davis became known as something of a provocateur, but by all accounts a tolerable one, attending conferences alongside invasion biology's more orthodox scholars and continuing to publish on less controversial topics, including the concept of "invasibility" and the ecology of native woodland ecosystems in the upper Midwest. His 2009 textbook *Invasion Biology*, which surveyed then-current research and controversies, received favorable reviews, despite its gentle suggestions that the subfield be dissolved, or at least reframed, as the study of how organisms—of any origin—change and expand their ranges (he would christen such a reorganized discipline "SPRED ecology," for "SPecies REDistribution").

Indeed, while he refuses to back down, Davis has also resisted a temptation that can plague heterodox thinkers, to become more extreme as the mainstream resists their ideas. He has never been a radical; the crux of his argument remains that, while some invasions do pose ecological and economic threats, regarding non-native species with automatic suspicion—or outright revulsion—is simply unscientific.

After 2011's "Don't judge species...," however, the temperature of the debate began to rise, as more critics of the field spoke out and more invasion biologists reacted. That brief commentary has by now been cited a not-too-shabby 1,200 times, which doesn't count mentions in the popular press, including by Marlene Condon in her fateful column. Of course, many of those citations have been in the service of fierce opposition—and no opponent has been more formidable than Daniel Simberloff.

The Heavyweight

"He's a friend," Simberloff says, in a Zoom call from his office at the University of Tennessee, Knoxville, when asked about Davis. "I don't have hard feelings about it. I don't think Mark does either. I just think he's wrong." (Simberloff, too, is evidently wrong on at least one point—Davis does have hard feelings. "I've cut off communication," he says, citing what he deems Simberloff's harsh treatment of some of the field's less well-established critics.)

It was Simberloff who wrote the response to "Don't judge species on their origins" that *Nature* subtitled "141 scientists object." He also led the recruitment of the 141 scientists—a notable number of cosigners, especially on fairly short notice, but Simberloff is by any measure a giant of invasion biology. His emails get prompt replies.

Simberloff earned his PhD in the 1960s at Harvard, under the renowned biologist E.O. Wilson. A gifted science communicator who went on to win the Pulitzer Prize for nonfiction (twice), Wilson was then working with the ecologist Robert MacArthur on what would become his first claim to fame, a theory of island biogeography. In 1966, Wilson and Simberloff set out to test the theory.

First, the pair selected a half dozen mangrove islets in the Florida Keys. They conducted a thorough census of all animals, which consisted solely of terrestrial arthropods, and then contracted with a Miami exterminator to tent and fumigate each islet in its entirety with the insecticide methyl bromide. (The highly toxic, odorless gas later turned out to deplete the ozone layer; it's now strictly regulated by the US EPA, which nevertheless allows its use on agricultural commodities to prevent invasive pests from entering the country.) The gas dispersed, leaving the vegetation untouched but the animals thoroughly dead. Over the next year, carrying an oar to fend off shark attacks in the shallow water, Simberloff conducted painstaking observations of the islets' recolonization by tiny booklice, grasshoppers, spiders, and ants as they arrived by wind or wing from other nearby islands.

The results of the Florida experiment—now a classic of ecology—were consistent with Wilson's idea that the number of species an island could sustain was inversely related to both its size and distance from potential sources of colonists—that is, smaller, more isolated places tend to have less biodiversity. This conceptual model has since been broadly applied to less literal islands, including habitats fragmented by human development, and while in recent decades it has been criticized as overly simplistic (including by Simberloff himself), it still informs scientists' understanding of island ecosystems' unique vulnerability.

Simberloff, whose Keys research became his doctoral dissertation, joined the faculty at Florida State University, where he and his colleagues—soon to be known as the "Tallahassee Mafia"—became involved in a dispute about the importance of competition in the formation of ecological communities. In 1975, Jared Diamond, the polymath now best known for his 1997 blockbuster *Guns, Germs, and Steel*, had published influential research based on studies of some islands around New Guinea. Diamond observed that bird species occupying the same ecological niche rarely co-occurred, a phenomenon he chalked up to competition between them. Skeptical of Diamond's seemingly too-neat conclusions, Simberloff and a graduate student tested the theory using computer simulations, concluding that the pattern Diamond had observed was actually random. Diamond and his allies struck back, and for a time the debate roiled ecology (both sides still have their partisans).

In 1997, Simberloff, by then known as a brilliant researcher who was not afraid to challenge ecological orthodoxy, moved to the University of Tennessee, which had recruited him for a professorship endowed by then Vice President Al Gore. It was there that he turned his focus to invasion biology, helping to inform Bill Clinton's 1999 executive order, editing the new journal *Biological Invasions* and in 2013 authoring the still widely read textbook *Invasive Species: What Everyone Needs to Know* (like Davis's book and this one, it was published by Oxford University Press). Today, Simberloff's is about the closest invasion biology has to a household name: *The New York Times*'s science desk routinely phones him for comment, which he is more than happy to give.

A bearlike man with expressive eyebrows and a penetrating gaze, Simberloff is the sort of old-school professor who doesn't suffer fools—and who may seek to expose one with a pop quiz. Tough grader though he might be, Simberloff's class on invasion biology has been popular with advanced undergraduates for decades, a record he attributes to the perennial appeal of the subject matter.

"They're reading articles about the Burmese python," he says. "Here in Knoxville, I see kudzu everywhere. They know about microstegium, and that there's huge problems. They're not stupid."

Invasion biology offers students concerned about the environment both an interesting set of problems to investigate and a way to feel that they're making a difference. Simberloff often connects them with extra-curricular "bush-bashes," projects to remove invasive plants in the area around campus, which provide his students with the satisfaction of putting theory into practice.

"Climate change, it doesn't seem there's anything you can do, or anything you do means so little," he notes. "Here you could do something."

Simberloff, who has helped shape invasion biology from the beginning, is bullish about its future as a science, if dubious about humanity's capacity to put its findings to good use.

"Progress is being made on understanding the scope, depth, and nature of the impacts of biological invasions," he says. "There's really been an enormous increase in our understanding of the range of impacts, the types of impacts, how idiosyncratic they are."

Despite the occasional skeptic, the scientific community is largely unified about the problems posed by invasive species worldwide, Simberloff asserts. In 2023, the United Nations–sponsored Intergovernmental Science-Policy Platform on Biodiversity and Ecosystem Services (IPBES) published its Thematic Assessment Report on Invasive Alien Species and Their Control, which was based on more than thirteen thousand scientific papers. The report estimated that humans had introduced more than thirty-seven thousand non-native species to biomes around the world, including more than 3,500 harmful invasives. Its authors were unequivocal in urging action against what they deemed "one of the five horsemen of the biodiversity apocalypse that is riding down harder and faster upon the world."

And there are signs that the war on invasive species is winnable, at least against some species, in some places. Simberloff is excited about recent progress toward eradicating non-native predators on islands. Wiping out mammals, which have caused most of the extinctions attributed to invasive species, is a bit more complicated than hiring a local exterminator to tent and fumigate, but it's not that different: There's a poison for almost anything you might want to kill. And a recent survey found hundreds of successful poisoning projects on islands around the world, including an 80 percent success rate against rats and 90 percent against carnivores like foxes and feral cats. As invasion biologists and land managers improve their methods, these efforts appear increasingly effective even on large islands with the complicating factor of a mobile human population.

Simberloff is less excited about what he calls the "sociopolitical" factors that prevent governments from acting quickly and efficiently upon the recommendations of scientists. While he remains committed to tamping down what he calls "bullshit," he sees dissent from within academia as less of a threat to invasion biology's mission than sclerotic politics and public disinterest in environmental protection.

Still, the vigor with which invasion biologists have responded to such criticism seems to belie Simberloff's contention that it's little more than a distraction. Indeed, a remarkable series of papers published in the late 2010s—and continuing on and off into the 2020s—suggests more may be at stake.

Fighting Words

"Remarkable is one way to put it," says Matthew Chew, the second author on "Don't judge species..." and Davis's frequent collaborator. "The papers were ad hominem by design: fatwahs declaring *personae non gratae*."

The tumult began with the January 2017 publication of an essay in the journal *Trends in Ecology & Evolution* called "The Rise of Invasive Species Denialism." Its authors, the New Zealand conservation biologist James Russell and Tim Blackburn, an invasion biologist at University College London, charged certain "high-profile media outlets," including *The New York Times* and *The Economist*, with challenging "the existing scientific consensus" on invasive species, despite an "overwhelming body of global evidence" that these organisms pose grave ecological threats. Worse, certain ecologists were "exploiting the fact that all scientific knowledge contains an element of uncertainty" as they "attempted to reframe, downplay, or even deny" the impact of invasive species. (Only Ken Thompson, Davis's collaborator and author of the 2014 popular science book *Where Do Camels Belong?*, was called out by name, but the rest knew who they were.)

As had happened after "Don't judge species on their origins," replies published in the subsequent issue of *Trends in Ecology & Evolution* were swift and fierce. Not eager to be equated with people who deny climate change and weave conspiracy theories about vaccines and the moon landing, Davis and Chew wrote to explain that they had only "called for appropriate nuance, complexity and objectivity" in a field that had "organized itself in the early 1980s to advocate one narrow, preliminary view of a complex, multifaceted story, and has stuck to that agenda ever since."

Replies to the replies poured in, spinning out into other journals. In *Biological Invasions*, two ecologists at McGill University attempted to quantify the "exponential growth of invasive species denialism," compiling a list of instances from both the popular and academic press and accusing the various "contrarians" of seeking fame as they sowed distrust in science itself. The alleged denialists denied the charges, which were followed by allegations—on both sides—of "logical fallacies," "cherry picking," and the construction of "straw persons."

Although many entries in the flurry of papers, letters, and essays paid lip service to the idea of "improving the debate," few participants ultimately saw the debate as having improved. Indeed, while the controversy ran its course largely out of the public eye (who, after all, cracks *BioScience* as light reading?) it did little to improve the discipline's reputation or effectiveness. While scientific controversies have been common since Copernicus's time, this episode of name-calling and impugning of motives comes off as shocking to the lay reader. After all, many of us are used to thinking of scientists as dispassionate truth-seekers.

"With the current plethora of niche disciplines, journals, and related social media, perhaps such things more readily take on a life of their own," says Chew, who personally bowed out of this specific argument after his first response. "I suspect most established, self-identified conservation biologists have chosen sides. Some of us (on both sides) are focusing on training incoming grad students."

Chew, who considers Simberloff a "frenemy," carries his own outsider status lightly—even gleefully. It may have something to do with his nontraditional path to academia; as a young man, Chew took a crack at a career in music and worked a series of odd jobs before starting college. Growing up in Scottsdale, Arizona, he had been fascinated by the riparian habitats along the Salt River and its tributaries ("That's where you go for anything other than cactus"), and as his thirties approached, Chew says, "it was looking increasingly inevitable that I needed to follow my interests in plants and animals."

Two degrees and a little more than a decade later, Chew was working as an ecologist for Arizona State Parks when a lifelong habit of speaking his mind got him into trouble. In 1988, the state had established Kartchner Caverns State Park to protect a pristine limestone cave system southeast of Tucson. Years of work followed to make the site accessible to visitors, but the $28 million project was not exactly a study in conservation best practices. Chew, who had previously developed the park's "Bat Management Plan," was shocked upon seeing the caves again shortly before they were opened to the public in 1999.

"I had been in there, just me and the bats," he says. "And I was astonished at the intervention that had gone on. Massive intervention. Lots of concrete, lots of changes. So that was a little distressing."

After saying as much to a National Public Radio reporter who had come to cover the final preparations, Chew found himself writing an essay for *The Boston Globe*, under the headline "A theme park grows underground." He was promptly fired, on the grounds that he had "sought to bring discredit and embarrassment to the state"—and that he had illegally used his "azstateparks.gov" email address to do so.

A whistleblower protection group, Public Employees for Environmental Responsibility, helped get Chew reinstated with back pay, but by then it felt too awkward to return. Chew's wife, the Arizona State University (ASU) ecologist Juliet Stromberg (also a coauthor of "Don't judge species . . .") encouraged him to pursue a PhD, and he wound up writing his dissertation about the pre-history of invasion biology, or how ideas about native and non-native organisms had developed in the decades and centuries before Charles Elton published what would come to be thought of as the subfield's founding document.

Since earning his doctorate in 2006, Chew has been a lecturer at ASU's School of Life Sciences, teaching and researching at the intersection of biology and human society. His long-running field class, Novel Ecosystems, brought undergraduates to habitats that fall somewhat short of pristine, including former agricultural land, urban riverbanks, vacant lots, and reservoirs. Chew encourages his students to see these places, with their mishmash of native and non-native species, as no less worthy of an ecologist's attention.

"People seem to prefer simple, familiar explanations to unfamiliar, complex ones. They like to scapegoat and demonize. They have short attention spans," Chew says, noting that the orthodox position—invasive species are bad—is well-suited to appeal to non-specialists. "But it becomes less tenable and more tiresome as one's knowledge increases, so 'the more you know' the more likely you will become to drift away from the received IB [invasion biology] view—unless IB is your identity and the source of your livelihood."

Intimations of financial interest are in fact common on both sides of the debate, with more orthodox invasion biologists accusing their critics of angling for whatever compensation allegedly comes with the attention of a click-seeking media (Chew and Davis have admittedly provided quotes to outlets from *Smithsonian* Magazine to local NPR affiliates, a habit that has yet to pay off). But while money can be a great motivator, identity—that of a good person, one who does the right thing, or of an expert, one who deserves the mantle of scientific authority—is often more powerful, especially when it seems to be under threat.

In a 2018 paper in the journal *Biology and Philosophy*, provocatively titled "Logical fallacies and invasion biology," Radu Guiaşu, a biologist at Toronto's York University, and Christopher Tindale, a philosopher at the University of Windsor, summed up the debate so far and offered a critique of the scientists accusing colleagues of "denialism." At base was a problem that may be inevitable in any field but which the authors argued was especially apparent in invasion biology.

"Science is touted as an objective discipline interested in discovery and descriptions," they wrote. "But those engaged in this discipline are human agents with all the foibles that attach themselves to such agents in whatever sphere they operate." Referencing the classic 1990 book *The Rhetoric of Science* by communications scholar Alan Gross, they noted how Gross saw "beneath the calm façade of the scientific paper a struggle to control meanings, attract followers, and establish authority."

In the wake of what might be termed the Denialism War, that façade seemed to erode almost completely. The revelation would have implications for important debates at the heart of invasion biology and for the all-too-human agents, from both the academy and the grassroots, hoping to carry it into the future.

No Room for Doubt

More than five years after her last Blue Ridge Naturalist column, Marlene Condon still sounds a little shell-shocked. Rheumatoid arthritis has shrunk her world, and she spends less time than she used to in the garden she has tended since 1986, on six acres of wooded hillside north of Crozet. But while she remains blacklisted at the *Crozet Gazette*, Condon still places occasional freelance essays at Charlottesville's *Daily Progress*, and she has not backed down on the issue of non-native plants.

"I get an awful lot of bad things said about me," Condon tells me. "I understand it's because a lot of people push this narrative. They just repeat these talking points. . . . I've really literally spent my life immersed in the natural world, so I know what goes on out there, and that's why I can see where people have things wrong. And I'm not afraid to try to correct it, because I feel like it's important."

That level of self-confidence can make for lonely going, and not just for those who veer from the mainstream. Susan Roth, the PRISM member who co-wrote the letter criticizing Condon, has found herself in awkward social situations when the topic of invasive species comes up.

"We were at a party once and there had been a fair amount of wine drunk," she says. Someone brought up Japanese honeysuckle (*Lonicera japonica*), an invasive vine that reached the United States in the 19th century. Roth and Hurley mentioned that it had been giving them trouble on their property. "And one woman just went nuts and started screaming about how good Japanese honeysuckle smelled, and she's not going to kill it, and just sort of attacking us for killing something that she loved."

Others have bristled when Roth and Hurley suggest people shouldn't garden with foreign cultivars or that herbicides have their place when used against invasive species. Roth even lost a friend over the invasive plants issue, after the woman expressed concern about the couple's use of herbicide on their land.

"She and her partner are very into nature and birds and wildflowers," Roth says. "They're intelligent people. They should have understood that we are restoring habitat."

The last time they met in person, Roth's friend handed her a copy of *Silent Spring*.

"That book is more than fifty years old," Roth told her. "We are not spraying DDT!"

The break was sad, but Roth, who is also chair of the Green County Democratic Committee, has come to understand the futility of arguing with someone who feels passionately. "I can't argue with people who are diehard Trump supporters," she says. "You're not going to change their minds. And you're not going to change certain people's minds about herbicides."

Blue Ridge PRISM's mission has recently expanded to a push for legislation and regulation, an area where it has found more success than Roth and Hurley's experience at cocktail parties might presage. In 2024, Virginia allocated $4.9 million to invasive species mitigation across the commonwealth, including through supporting other regional PRISMs modeled after the Blue Ridge organization.

Still, the real, day-to-day work happens out in the woods.

Roth and Hurley are getting older, and bushwhacking has taken its toll. Hurley shows off forearms scratched and scarred by multiflora rose thorns, but he avows that such wounds won't deter him. In fact, he decided to skip a planned summer trip to Italy with Roth so he could remain on the battlefield.

"The consequences to me of walking away, of leaving this property for two and a half weeks during this peak growing season are too great," Hurley says, sounding both regretful and determined. "At the end of the day, this is going to be a question of stamina and resources. One of us is going to outlast the other, the seed bank of the invader—or us. And if we lose that battle, then all those other gains that we've described to you—the expression of native species on the landscape, the reconstitution of a native plant community that has much richer ecological value for the native fauna—that will all be lost."

Interlude

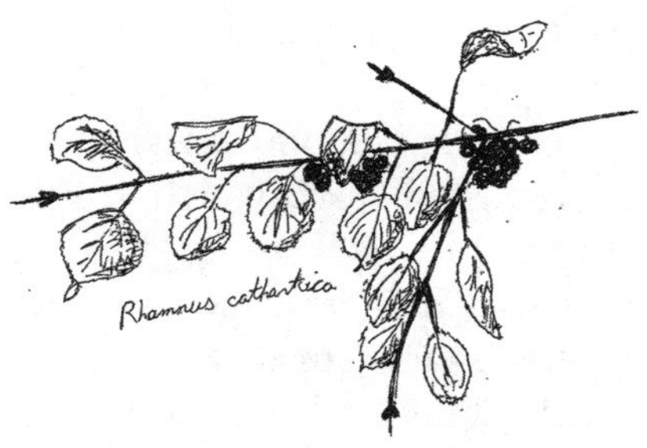

BUSH BASH

It's a weekday morning in mid-September, and I've joined a handful of volunteers at a county park 10 miles north of Ann Arbor to tackle a stand of common buckthorn (*Rhamnus cathartica*), which is among the most ubiquitous—and least-loved—invasive plants in our area.

The maples are tinged here and there with orange, but this year summer is taking the long way out. By 10 a.m., when our shift starts, the morning mist has burned off, and beneath a cloudless sky the air lies hot and still, the lake where my kids swam last month smooth as glass. A pair of sandhill cranes grazes nonchalantly beside an empty lifeguard's tower, while mute swans glide along the opposite shore.

Armed with two-foot loppers and folding hand saws, we head across the gravel parking lot, over a boardwalk, and into the woods, to a triangle of forest wedged between a hiking trail and the shore of a large wetland. Dominating the understory beneath a few towering red oaks is about as close to a buckthorn monoculture as you'll find anywhere, although at first it's hard to pick out the individual trees, with their plain ovate leaves and non-descript bark. The giveaway is the toxic black berries glistening among the greenery.

We get to work cutting down trees, which range in size from saplings to a mature specimen, perhaps 15 feet tall with a trunk six inches in diameter, that a park staffer guesses has been there for at least ten years—the mother tree, perhaps, for the whole patch. A saw blade breaks on its lichen-covered trunk before the tree finally falls.

Stumps must be dabbed with herbicide to prevent the trees from growing back, so the Parks Department supplies us with Buckthorn Blasters® (available from the online shop of the North American Invasive Species Management Association, which also sells "Leave Only Footprints" hoodies and phone cases). These small, refillable bottles contain diluted glyphosate, which is dyed a vivid blue so that we can keep track of which stumps have been treated. The bottles may be spill-proof, but nitrile gloves are recommended.

This patch of woods is fairly small and not too far in any direction from roads, park facilities, and neighborhoods, but even setting aside the buckthorn there is plenty of life here. As we work, the toothy scratching of saws and the snapping of branches is accompanied by the plunks of falling acorns, the chirps of chipmunks, the drone of tree crickets, and, thrillingly, the call of a pileated woodpecker. Spiders have set up shop in the undergrowth, and mosquitos harass us unceasingly. In the upper branches of one toppled buckthorn, a volunteer finds the abandoned nest of a red-eyed vireo, an intricate bowl of dried grass glued securely into a frame of twigs with spider silk.

As the buckthorn falls, we dismember the larger trees and stack debris just off the trail (park staff will take care of it with the next prescribed burn). Fruit-bearing branches, however, require special care, because any seeds that make it to the soil might sprout next spring. We impound these biohazards on a heavyweight plastic tarp, creating a chest-high pile that requires four of us to drag back to the parking area. Later, a utility vehicle will haul it to the dump.

Sweaty, bug-bitten, and tired, I look around at the end of my shift. There is certainly a lot less buckthorn. In its absence, other plant species are visible: a few shagbark hickory saplings, riverbank grapevines, plenty of poison ivy, and some sickly looking ash trees. Among the blue-painted stumps sits a mossy log, the remnant of a fallen giant whose demise, years ago, must have opened the canopy and provided the buckthorn with its opportunity to take over. We've opened the canopy again today, and while the patch of woods, the bare earth pocked by blue-painted stumps, now looks—frankly—awful, we hope that we've given the shagbarks what they need to thrive and that perhaps a few acorns now have a fighting chance to become mighty oaks.

Either way, the volunteers will be back. And so, almost inevitably, will the buckthorn.

5

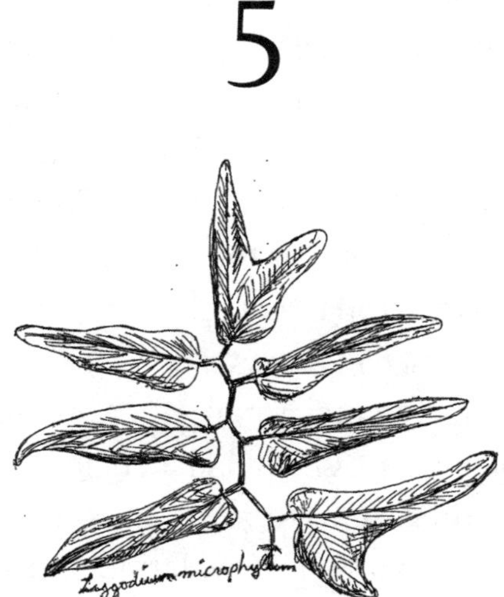

WEEDS IN THE RIVER OF GRASS

Brendon Hession pilots his square-jawed airboat deftly through Loxahatchee National Wildlife Refuge, a 226-square-mile remnant of the historic Everglades just west of Palm Beach. An invasive species biologist with the South Florida Water Management District (SFWMD), Hession knows the territory well—or as well as any human can know this dizzying maze of sawgrass ridges and sloughs, scattered with islands and gator holes. As the boat skims noisily along, it surprises bitterns, egrets, herons—and the occasional good-size alligator, which dives hastily beneath the tea-brown surface.

Several miles into the refuge, we reach a pair of airboats parked among pond lilies, and Hession cuts the motor. For a moment, the silence of this vast, prehistoric wetland descends on us—then, gradually, we hear laborers making their way out of the undergrowth on a nearby island, chatting to one another in Spanish as they sploosh through thigh-high water. They wear chest waders, sunglasses, hats, and bright orange rubber gloves. Each carries a four-gallon backpack herbicide sprayer.

The men are contractors for the water management district, and their main target today is Old World climbing fern, also known as *Lygodium microphyllum*, or "lygo" for short. It's part of a genus found throughout equatorial Africa and Asia, where local people use its arrowhead-shaped leaflets in a poultice for skin ailments and its strong stems to weave baskets, or the traditional wide-brimmed hats worn by rice farmers in the Philippines. Lygo also makes an attractive houseplant, which is how it wound up in South Florida sometime before 1965.

It wasn't long before this pretty fern escaped captivity.

Spread by microscopic spores that can travel miles on the wind, lygodium takes root easily and is hard to kill. It thrives in standing water as well as it does in moist soil, tolerates both bright sun and deep shade, and can even survive a full washer and dryer cycle.

"Anyone who works with this stuff, just check the spot where their dryer vents," Hession says. "I have a bunch that I pluck out every once in a while near mine. And you'll find lygodium coming up where your AC drips."

In a suburban yard, it's an annoying weed, but in the Everglades lygodium has spread at a remarkable—some say disturbing—rate.

In the 1990s, soon after lygo reached Loxahatchee, *The Washington Post*'s Joel Achenbach compared it to "the Blob." The prize-winning reporter wrote at a time when invasive species panic was sweeping the media, but while his pulp science fiction references may have been a touch overblown, a quarter century later it's easy to see what he meant: start looking and lygodium is almost everywhere.

It's especially vigorous on the small knobs of land known as "tree islands" that are scattered throughout the Everglades. Ranging in size from less than a quarter-acre to nearly 20, these low-slung islands are thusly named because they offer enough dry—or dryish—ground for endemic trees like swamp bay (*Persea palustris*) and pond apple (*Annona glabra*) to take root. Although tree islands are not the most common habitat in the Everglades, they're a key part of the overall system, serving as rookeries for wading birds and nesting sites for reptiles. They also play an important role in regulating soil phosphorus, influencing the flora of the surrounding marsh. But the same elements that make them crucial to native plants and animals also make them hospitable to lygodium.

"You've got a lot of structure here for it to clamber onto. It can just totally dominate the canopy," says LeRoy Rodgers, SFWMD's section administrator for vegetation management and Hession's boss. "It's perfect habitat for this plant."

Flocking to Florida

The Everglades has more than its share of problematic non-native species, from the ubiquitous hydrilla (*Hydrilla verticillata*), an aquatic plant native to the Old World but now found everywhere except Antarctica, to the attention-grabbing Burmese python. But invasion is only one in a long line of manmade threats facing this vast wetland. Indeed, a cursory look at the past century and a half of South Florida history suggests that the Glades may have been doomed from the minute land speculators first took note of Florida's pleasant winters.

Early efforts to drain the swamp with ditches and canals date to the 1880s, and by the 1920s real estate and agriculture were steadily chipping away at its edges. It was not easy going, however, and despite kicking off the rapid growth of Miami and other coastal cities, a frenzy of land speculation devolved into an enormous bubble, which ended in 1926 when a hurricane swept across the peninsula, killing hundreds, causing some $1.7 billion in damage (adjusted for inflation), and making painfully clear how difficult it would be to create whole cities out of swampland.

The bust contributed to the misery of the Great Depression, but ever-optimistic Americans declined to give up on the Florida dream. After World War II, developers began carving off chunks of the Everglades again, dredging canals and filling in bays to construct subdivisions for returning veterans. Meanwhile, shifty "swamp peddlers" set up shop again, swindling retirees out of billions while leaving a grid of abandoned streets running through cypress stands and wet prairies.

The SFWMD and other government agencies are now working to undo some of this legacy. In 2000, Congress authorized the Comprehensive Everglades Restoration Plan, or CERP, as a partnership between Florida and the federal government. It's estimated that work will continue through 2050 and cost nearly $28 billion in 2024 dollars. While the area's complex natural hydrology is irreparably altered, the project aims to restore much of the sheet flow that once carried fresh water from north to south, supporting habitat and recharging the aquifers on which humans have come to depend.

Still, today just half of the historic Everglades remains, and development pressure hasn't ended. While much of the wetland is now protected in a network of state and federal lands, including the Loxahatchee Refuge (established in 1951) and the 2,357-square-mile Everglades National Park (1947), over the past forty years Florida's population has doubled to more than twenty-three million people, all of whom need to live somewhere. A labyrinth of strip malls and subdivisions now runs almost flush with the broad canals that demarcate the Everglades' eastern flank, while the few remaining parcels of dry land bristle with excavators and cranes. An analysis by the University of Florida's GeoPlan Center warns sprawl could cover 70 percent of the state by 2070.

Despite everything, today's Everglades remain a hotspot for biodiversity. They are home not only to the iconic American alligator (*Alligator mississippiensis*) but also to more than fifty species of lizards, snakes, tortoises, and turtles. Some 360 species of birds live here or pass through on their annual migrations, including such charismatic crowd-pleasers as the American flamingo (*Phoenicopterus ruber*) and roseate spoonbill (*Platalea ajaja*). Endangered manatees (*Trichechus manatus*) find refuge here, along with the last few dozen Florida panthers (*Puma concolor coryi*).

Over the centuries, the Everglades has also provided sanctuary for humans. During the Seminole Wars of the early 19th century, Indigenous people forced out of their historic homes in what had recently become the state of Georgia hid from the US military in a wet wilderness that white Americans would grimly describe as "impenetrable." By camping out on tree islands, a few hundred Native Americans were able to avoid forced removal west of the Mississippi; today, their descendants, the Miccosukee and Seminole tribes of Florida, hold tree islands as a sacred part of their heritage. Some are reserved as cemeteries, while others serve ceremonial purposes, including as sites for the annual Green Corn Dance, a secretive ritual that few non-native people have ever witnessed.

Now, with Old World climbing fern taking hold, tree islands have become battlegrounds in the fight to save the Everglades as Floridians know it.

"I'm not exaggerating when I say I think this is probably one of the most challenging invasive plant management projects in the country," Rodgers says. "It could take a decade to get this thing under control."

Ridding an island of climbing fern without also killing the native vegetation is extremely labor-intensive. Workers will often begin by hacking through vines at the base, an approach Rodgers and Hession call "poodle-cutting," because the result resembles a show poodle's distinctive hairdo. Islands that have been poodle-cut stand out, with brown puffs of dead lygodium slowly rotting at the top of whitened snags—all that remains of the trees the fern had used as scaffolding. Once the lygo tops are severed, crews spot treat surviving stems at ground level with the herbicide triclopyr, which targets broad-leafed plants.

Of course, all that only makes sense when the lygodium hasn't taken over completely. The places where it has stand out, too. While a classic tree island has a neat dome shape against the horizon, a heavy mat of lime-green lygodium can collapse the canopy like a punctured balloon, leaving dead trees wrapped with layers of fern. The tendrils continue to climb up and over themselves, curling and twisting and falling over until the island resembles a series of neglected topiaries: here a bent giraffe, there a trumpeting elephant. For management purposes, this is end-stage lygodium.

"When you start getting to where the backpack crews are going to have to swing a wand over 90 percent of the area, everything is going to get herbicide on it no matter what," Hession explains. "So, for an island like this, we can hit it with a helicopter."

Some of the native plants, especially myrtles, native ferns, and sawgrass, will also succumb after an aerial application, Hession acknowledges, but others, including the big bays and hollies, have waxy leaves that the chemical can't penetrate so easily. Of course, some of the lygodium will also, inevitably, survive, but there will be less of it—hopefully enough that a ground crew can take it out with spot treatments a few months later. Meanwhile, if everything goes right, native plants will begin to bounce back once the initial chemical application wears off (of course, lygo spores are equally capable of recolonizing, should the wind carry them that way).

A lanky former marine who spent two years as a machine gunner in Afghanistan, Hession carries a tablet computer with a satellite view of the refuge, divided into numbered, one-kilometer cells. The islands normally appear dark green when viewed from above, but color-coded overlays tell him how serious the invasive plant situation is on each, when he last assessed its condition, and when it was most recently hit with chainsaws or herbicide by a crew of contractors.

I suggest that the map is like his battle plan, then realize with chagrin that talking about exotic plants that way might seem a little glib when compared with the war in Afghanistan. But Hession says he doesn't mind—he thinks in those terms, too. The first priority is to push back the enemy threat. Then, the SFWMD and its allies work to keep the threat under control, finally allowing the native ecosystem to peacefully reassert itself.

"Clear, hold, build," Hession says with a wry smile, quoting the US counter-insurgency strategy.

Collateral Damage

While those on the front lines of conservation work strive to use herbicides judiciously, they also admit that relying on these chemicals—most often the broad-spectrum glyphosate, best known as the active ingredient in the commercial weed killer Roundup, or increasingly triclopyr, which has the advantage of being more precisely targeted—is not exactly a feel-good approach, even when it seems to be the only option.

It's an inescapable irony that habitat restoration so often involves wholesale destruction. And while shooting or trapping animals may be more obviously unsettling, removing invasive plants can also be remarkably violent. Tearing out a stand of shrubs, like autumn olive, may require the sort of heavy equipment typically used in road work: think skid-steer loaders, bulldozers, and heavy-duty tractors. Tough root-raking attachments pull up dense patches of kudzu, which must then be piled up and burned lest it resprout. Large trees, like the white pines (*Pinus strobus*) and Australian acacias (genus *Acacia*) introduced to South Africa's sensitive fynbos ecosystem, are often dispatched by girdling. That process involves gouging out a ring of bark near the base of the trunk and cutting into the phloem just beneath it, stopping the flow of water and nutrients so that the tree eventually starves.

Once an area is cleared using one of these methods, deliberate reseeding of native species can spark the recovery of natural habitat—but meanwhile, it's tough to avoid collateral damage, whether to streambeds and soil structure (which contributes to erosion) or to any civilian plants and animals that might get in the way. And while a scorched-earth strategy may look effective, physical removal on its own is rarely enough to eliminate all invasive plants from an area. In fact, the disturbed soil can actually offer non-native species a new toehold.

And that's where herbicides often come in.

"Herbicides have an interesting history within the world of invasive plant management," says Stephen Enloe, a professor at the University of Florida's Center for Aquatic and Invasive Plants whose research focuses on chemicals for invasive plant control. "Early on, a lot of folks who were interested in management and restoration had an anti-herbicide mentality: 'We don't want to add chemicals to the environment, we're trying to get away from things like that.' And so they attempted virtually every non-chemical strategy for invasive plant management—and it turned out a lot of things are just relatively ineffective or economically sort of impossible."

Hand-pulling, seemingly the least destructive approach, does little, Enloe says, especially since many invasive plants have adapted to such disruption by tearing easily, leaving root fragments in the soil to regenerate (think about how hard it can be to dig a dandelion's taproot out of your lawn).

"That tremendous labor just results in removal of top growth, and they then regenerate very rapidly, and so you're sort of back to square one," Enloe says. "And with digging things out, often the cure is worse than the disease, because it's a complete ecosystem disruption when you're coming in with bulldozers."

Tarping, or "solarization," in which plastic sheets are laid over invasive plants to effectively cook them over several weeks, is another common method with unclear secondary effects,

Enloe says. "You're laying down massive amounts of plastic across an ecosystem. What's the fate of that plastic in the environment? Is it in itself a problem?"

With few easy answers, herbicide has often proved the best option, Enloe says, and over time scientists and managers who were initially hesitant to rely on chemicals have come around. A survey of land managers conducted by the California Invasive Plant Council in 2014, for example, found that 72 percent "frequently" or "always" used herbicides to control invasive plants, while only 6 percent claimed never to use them.

"We've seen this occur over the last twenty years," Enloe says. "Some folks who would not use herbicides before will definitely lean into herbicide use, when they can do it in as selective a manner as possible. It's not my goal to blast everything out there. Every herbicide study I conduct, the primary focus is, 'What's the lowest possible concentration I can use to be effective, and how do I facilitate recovery of native plant communities where I am selectively controlling that invasive plant?'"

As land managers came around on herbicides, glyphosate soon emerged as the most popular chemical for invasive plant control.

In many ways, glyphosate is a miracle. A synthetic compound, it works against nearly any plant by inhibiting a certain enzyme used to build proteins. Once exposed through their leaves or stems, plants die quickly, while any leftover glyphosate binds tightly to soil minerals, where most of it is broken down by naturally occurring bacteria within a few days. Traces of the chemical may linger in the soil for six months, but plants can still recolonize a treated area fairly quickly, which means a farmer who sprays a field with glyphosate doesn't have to wait long before planting a crop there.

The agribusiness giant Monsanto patented glyphosate in 1971; three years later, when Roundup (coined because, unlike previous generations of herbicide, it will "round up" all the weeds in a field) reached the market, it was an instant hit with farmers, who watched their yields increase dramatically. Roundup soon became the leading herbicide worldwide, and its use expanded further when Monsanto's first patent expired outside the United States in 1991, opening up the field to generic competitors.

To make up for its lost patent, in 1996 Monsanto introduced "Roundup Ready" crops— first soybeans, later corn, canola, cotton, and sugar beets—which were genetically modified to be immune to glyphosate. That meant the chemical could be applied to a field again and again over the growing season, killing recurrent weeds without harming the crops themselves. Farmers also began using glyphosate to "dry out" non-resistant row crops like wheat and oats, allowing for an earlier harvest. Between 1974 and 2014, glyphosate use grew 300-fold, and today more than 800,000 tons of it, in more than 750 commercially available formulations, is applied to agricultural land every year. Another 10,500 tons are used on parks, road verges, golf courses, cemeteries, and gardens, and in habitat restoration.

With such a successful flagship product as a lure, it's no wonder Bayer, the international pharmaceutical and chemical conglomerate, paid $66 billion to acquire Monsanto in 2018. But the acquisition also brought trouble. Over the past few years, Bayer has faced a barrage of more than 125,000 lawsuits from cancer patients alleging links between Roundup and their diagnoses. In 2019, a school groundskeeper who had been accidentally drenched with glyphosate-based herbicide and later developed terminal non-Hodgkin lymphoma won $289 million in punitive damages (an amount that was later reduced by a judge). Since acquiring Monsanto, Bayer has set aside $10 billion to settle lawsuits filed by lymphoma patients. In 2023, it removed glyphosate from its residential lawn and garden products, although the chemical remains available for professional use.

The plaintiffs' cases in the glyphosate lawsuits are bolstered by the International Agency for Research on Cancer, part of the World Health Organization, which in 2015 found that glyphosate was a "probable human carcinogen." Meanwhile, a growing body of research indicates that, at least in vitro, glyphosate affects gene expression, hormone activity, and metabolism in human cells, bolstering evidence that the chemical could cause cancer. Animal studies have found that high or chronic doses of glyphosate formulas can damage lungs, livers, and kidneys. Researchers in Europe have also shown that glyphosate harms earthworms and the microorganisms in soil that help plants take up phosphorus and nitrogen.

In general, however, the science has been frustratingly equivocal about the impact of society's dependence on glyphosate.

"On the one hand some authors report that glyphosate is less toxic than table salt," note the authors of a recent metastudy published in the journal *Encyclopedia of the Anthropocene*, who surveyed more than a decade of research into the question of glyphosate's impact on human health and the environment. "On the other hand, other authors report that glyphosate is involved in the etiology of many diseases. . . . This extreme discrepancy in the scientific community has brought confusion at the regulatory level."

Bayer's legal settlements have included no admission of wrongdoing, and the company continues to assert that Roundup poses no threat to human health. Its position has been repeatedly backed up by the US EPA, which regularly reviews available data and maintains that glyphosate is "not likely to be carcinogenic to humans," nor are there any other risks to human health when products containing the herbicide are "used according to label directions." (That presumably excludes dousing people in the chemical from head to toe, as sometimes happens in workplace accidents.) The agency has declared that it's safe to eat food with trace amounts of glyphosate—which, incidentally, is unavoidable given the chemical's ubiquity—and for children to play in areas that have been treated with it. The EPA does see a potential risk to bees (the agency says it requires more data to rule that out), but it has found no other reason to be concerned about glyphosate's impact on animals or their habitats.

Still, commercials advertising class action lawsuits continue to solicit glyphosate plaintiffs, and even scientists who recognize the importance of herbicide to habitat restoration have their qualms about its overuse. For one thing, certain agricultural weeds have evolved resistance to glyphosate, including kochia (*Kochia scoparia*), an introduced species in the United States, where it's also known as "fireweed." Meanwhile, other herbicides have their own possible downsides. In certain forms, triclopyr, which was developed by Dow Chemical fifty years ago, is highly toxic to fish. It can settle in mammals' fat, ovaries, livers, and kidneys, and studies have tied it to reproductive changes. The EPA has determined triclopyr is safe to use but notes that studies are still needed to determine for sure that it's not a human carcinogen.

For Enloe, the suspicion and uncertainty surrounding herbicides has only made his job harder. In addition to his research and teaching duties, Enloe is also an invasive plant extension specialist, which means he communicates the latest science to other agencies, nonprofits, land managers, and volunteers involved with invasive species control.

"I've had to shift literally about 25 percent of my program into just talking about glyphosate all over the place," Enloe says, grimacing. "We talk through the history of glyphosate, this carcinogenicity thing. We talk about other environmental issues, because independent of the EPA data, a lot of academics and eco-toxicologists are digging into glyphosate right now. The quality of that science is often very high and occasionally very low, and so there's a mixed bag of data out there."

Breakthrough technology is a big reason why glyphosate is getting more scrutiny—today, scientists have the ability to detect a chemical's presence all the way down to parts per trillion, and they have found glyphosate in everything from bagels to baby formula and in a majority of urine samples from people around the world, even in remote villages (whether that's a problem for these people remains uncertain). At the University of Florida, researchers recently discovered glyphosate in the blood plasma of manatees living in the spring-fed Crystal River, which is generally considered as pristine as you can get in the Sunshine State.

"The Crystal River of all places, where glyphosate is not utilized!" Enloe says. "How did it get there?"

Even Enloe himself, who has studied herbicides for years, has mixed feelings about the chemical. On the one hand, it's such a useful tool. On the other, he says, "I find it very unacceptable that glyphosate is present in the blood plasma of manatees."

Seeking New Recruits

Over the course of the 20th century, Floridians went from resenting the Everglades as an inhospitable waste to embracing this extraordinary wetland, with its gentle manatees and fearsome alligators alike, as an ecological treasure that makes their state unique. Meanwhile, official attitudes toward non-native species went in the opposite direction. Not too long ago, after all, David Fairchild established the USDA's experimental garden at the swamp's edge and dispatched Uncle Sam's Plant Hunters to seek out new species for cultivation there.

"This is the beginning of the 20th century, where we were really concerned with plant diseases, but we weren't really concerned with pest plants—it hadn't really hit the forefront that these could grow out of control and be ecologically damaging," says Melissa Smith, a research ecologist with today's (more enlightened) USDA. When the Everglades were considered a wasteland, no one worried—or even noticed—that Florida's climate offered imported organisms a favorable habitat in which to spread unchecked. "It was out-of-sight, out-of-mind," says Smith, whose tanned arms are laced with botanical tattoos, as she shows me around the USDA's modern-day Invasive Plant Research Laboratory in Davie, a suburb of Fort Lauderdale.

The lab's origins date to 1953; the following year, Fairchild died at his villa in Miami's Coconut Grove neighborhood. The vaunted plant explorer, long retired from government service, had spent the last decades of his life there cultivating a personal version of the USDA's plant introduction garden, which he and his wife called "The Kampong," after a Malay word for "village." Lush, green, and haunted by feral peacocks (*Pavo cristatus*, native to India), the 9-acre oasis—today open to the public by appointment—retains an obvious appeal. An epic banyan (*Ficus benghalensis*) guards its gate, while a collection of palms lines the walk to a point overlooking the blue waters of Biscayne Bay. Fairchild once wrote, "I have always claimed that the planting of trees from other lands has a romance about it that nothing else has." This place seems designed to drive home the point.

Thirty miles north, and with an aesthetic best described as "functional," the USDA lab is somewhat less romantic. Still, Fairchild himself may have felt at home here. Like his Kampong, the low-slung laboratory building is also surrounded by tropical plants, including aquatic weeds like water hyacinth (*Eichhornia crassipes*) growing in a grid of oblong concrete boxes. ("These are actually coffin vaults," Smith says. "I know! But they work great.") Nearby, horticulturalists have created an education garden to contrast Florida's natives with its most successful exotics.

"Here's melaleuca," Smith says, of a tree with peeling gray bark that grows next to a small pond (at our approach, several non-native green iguanas—*Iguana iguana*—plop into the water like falling fruit). "You can see why they call it paperbark. And it's just chock-a-block full of volatile chemicals, very flammable." This was one of Fairchild's introductions, she notes. "I'd be remiss if I didn't say we're kind of undoing his work, in some ways."

Melaleuca quinquenervia was not solely Fairchild's fault, of course. Records show that many people introduced the so-called Cajeput tree to Florida in the years before and after 1900, with and without the USDA's help. Later, the Army Corps of Engineers used this fast-growing Australian import to stabilize levees along Lake Okeechobee in the 1930s. Their effort was assisted by freelancers who hoped water-hungry melaleuca would help dry up the swamp and make it profitable. But it would be the USDA that eventually turned the tide once melaleuca spread out of control—this time with the help of another kind of introduced species.

According to what's known as the "enemy release hypothesis," when a species is transplanted to a new range, it no longer has to contend with the predators, parasites, or pathogens with which it co-evolved and which previously kept its numbers in check.

Studies have shown, for example, that plants have an average of three parasites where they are introduced, compared with sixteen in their native range. Though only fully fleshed out by an influential *Trends in Ecology & Evolution* paper in 2006, the idea dates back to Darwin and was cited favorably by Elton.

This is an area of science where practice has long been ahead of theory, as for more than fifty years seeking out and releasing natural enemies, usually referred to as "biological controls," has been a major project of invasion biology. (Recall Bernd Blossey, the Cornell professor who studied garlic mustard and who got his start importing insect enemies of purple loosestrife.) In the 1960s, the USDA partnered with the Army Corps of Engineers to seek out and test biological controls for aquatic weeds, which in addition to their ecological impact also threaten agriculture and human communities by complicating irrigation and water management. With colleagues at other USDA Agricultural Research Service labs, the scientists in Davie investigated flea beetles, moths, and thrips, releasing promising species and watching as they chewed through invasive alligatorweed (*Alternanthera philoxeroides*) across the American South.

By the 1990s, with USDA researchers declaring alligatorweed's retreat "the World's First Aquatic Weed Success Story," melaleuca had become a top research priority for the Davie lab, and researchers traveled to Australia to find arthropods that could stunt the tree's growth or hinder its spread. First, they found a brown bug called the melaleuca leaf weevil (*Oxyops vitiosa*), which feeds on new melaleuca growth; some eight thousand were released in the Everglades in 1997. They were followed in 2002 by the melaleuca psyllid (*Boreioglycaspis melaleucae*), and then the melaleuca gall midge (*Lophodiplosis trifida*) in 2008. Working together, these insects can cut plant growth by half and reduce seed production by up to 99 percent.

"Melaleuca is one of our biggest successes," Smith says, noting that across the Everglades acreage covered by the tree is down 96 percent since the biocontrol program began. "Which is not to say it doesn't pop up and there aren't some very specific areas that are persistent, but when we look at things on a landscape scale, it's really effective."

Anyone who has followed the invasive species narrative so far might raise an eyebrow at the idea that introducing yet more new organisms into the mix won't have unintended consequences. And indeed, humans who hope to recruit non-native enemies to the fight against a pesky organism have made high-profile mistakes before. The most notorious is the cane toad (*Rhinella marina*): Brought to Australia in 1935 to eat beetles plaguing sugarcane plantations, the toxic Latin American amphibian is now charged with causing widespread ecological disruption as it preys on smaller native fauna and poisons larger animals that attempt to eat it. (Cane toads were also introduced in Florida, where—like so many other species—they quickly made themselves at home in the Everglades.)

Unlike the cane toad fiasco, however, modern biological control efforts are not haphazard affairs, and a generalist predator like the toad is actually anathema to the concept. Releasing a potential biocontrol agent into the wild comes only after years of careful study, to ensure it will prove harmless to all but its intended target.

Under strict conditions, scientists must pit potential biocontrol agents against native species that are closely related to an invasive target—and even distantly related, just in case. They observe interactions between species for multiple generations and investigate any possible adverse impact, however unlikely it might seem. They make sure caterpillars won't poison native birds and that introduced species won't serve as reservoirs for disease. If there is any indication that a potential biocontrol could itself cause ecological or economic harm, that line of research is immediately abandoned and all remaining individuals killed by freezing, followed by careful disposal of their remains.

"Modern biological control has a 99.6 percent safety record," Smith says. "And that last 0.4 percent is just from one weevil."

That insect, the Canada thistle bud weevil (*Larinus planus*), was released by the US Forestry Service in Colorado in 1992 to combat Canada thistle (*Cirsium arvense*; despite the common name, it's originally from Europe). Several years afterward, the bug was found laying eggs on native thistles, including several rare species. Federal regulations have since been tightened, Smith says, adding that today "that weevil never would've been released."

Indeed, Smith's lab, a Biosafety Level 2 facility, observes stringent quarantine rules. To get a look at the biocontrols she and her team are working on now, I follow her through a steel door into a vestibule, where we don long coats and hair coverings to guard against any stowaways when we leave. Then it's through another door into a tiny chamber, where we wait for that door to close before opening the door into yet a third chamber.

"This is a three-bay airlock," Smith explains. "Feel the negative pressure? So, the idea is to suck insects back into the lab."

The airlock is a little claustrophobic, but we soon emerge into a spacious, light-filled lab. Along one side are attached greenhouses where the researchers tend invasive plants in pots and pit them against whole hordes of hungry insects. Through another door is an isolation room, the first stop for potential biocontrols after they arrive from overseas in a scientist's hand luggage (always nestled within a requisite "three layers of containment"). Here, the newly arrived arthropods are shepherded through their first stateside generations to ensure they are themselves free of parasites and diseases.

"These just came in from either South Korea or Japan," Smith says, pointing to a series of small jars that seem to contain nothing but skinny twigs. The invisible inhabitants are stem borers who live on cogongrass (*Imperata cylindrica*), a perennial listed as a federal noxious weed in the Southeast United States. This particular insect is new to science, and until a taxonomist comes up with a formal name the team has dubbed it "Green Goblin."

"Stem borers are a whole new nightmare to rear, because you just have to trust that they're there, but frass is usually a good indication," Smith says, pointing to a few tiny pellets of poop. "Stem borers are very tidy. They'll clean out their little burrow on a daily basis."

Nearby, a technician carefully transfers moth pupae into individual cups with a pair of tweezers, while behind her newly emerged adults swarm around a bouquet of lygodium inside a large plexiglass habitat. In other cases, caterpillars—tiny and black, brown and camouflaged, or fat and green—munch away on fresh leaves.

"This one we're pretty excited about because it's a relatively large caterpillar, so it eats a lot," Smith says, of a species known as *Callopistria exotica*, which metamorphoses from a very hungry caterpillar into a fuzzy brown moth. "I was in Hong Kong with one of our Australian colleagues and I saw lygodium that was just totally denuded. . . . I mean, you can really see the damage! They're also kind of cute, which is always helpful for us."

These loveable lepidoptera are slated to be released soon, joining previous conscripts in the war on Old World climbing fern, the brown lygodium moth (*Neomusotima conspurcatalis*), and the mite *Floracarus perrepae*, an herbivorous arachnid that forms damaging galls on lygodium leaflets. In 2010, the USDA partnered with the SFWMD and the US Army Corps of Engineers to build and operate a 2,700-square-foot annex—billed at the time as a "bug nursery"—at the Davie lab. In recent years, moths have been released by the hundreds of thousands and mites by the millions at locations throughout the Everglades. They have since established spreading, self-sustaining populations.

"Unlike herbicide, biological controls are slow," Smith says. "This works on a biological time scale, rather than 'Spray it and it dies.' And, of course, an effective biological control agent does have an upfront development cost. But they are free once they're released."

The melaleuca leaf weevil has been thriving on its own in the Everglades for nearly three decades, she notes. And though herbicide is still deployed against the tree in some cases, thanks to the weevil and its friends, chemicals are now used less and less often. The SFWMD, for example, has cut its use by two-thirds on a per-acre basis over the past fifteen years, LeRoy Rodgers reports.

Air potato leaf beetles (genus *Lilioceris*) represent another smashing success, this time against *Dioscorea bulbifera*, a fast-growing yam relative from tropical Africa and Asia that, like Old World climbing fern, can twine around and smother Florida's native vegetation.

"This cute little red beetle has absolutely eliminated the need for treating air potato with herbicide," Smith says. "People are ecstatic about that. When biocontrol works, it's astounding. It really changes things, and native species come back."

Relying more on biology and less on chemistry requires not only accepting a longer timeline—one that includes years of careful testing as well as time to take effect in the wild—but also abandoning a way of thinking about invasive species that has dominated ecology and conservation for generations.

"We've been saying, 'Kill it, kill it, kill it!' for years," Smith says. "It's a little tough to shift that paradigm."

We're fooling ourselves if we imagine invasive species can be eliminated, this new thinking goes, but it may be possible to better integrate them into the ecosystems where we've introduced them, tamping them down to a level where they don't harm preexisting species. After all, the plants Smith's lab targets weren't considered a problem in their native range. If invasion only happens when a species is freed from the environmental challenges that once kept it in balance, the solution may lie in making its new home just a little bit more like its old home.

Environmental Engineering

Florida's five water management districts are regional authorities that work to prevent floods, conserve wetlands, and ensure adequate water for human use. These are especially crucial jobs in a state where the boundary between intense human development and utter wilderness is especially thin, and where humans' dependence on nature's good behavior can be starkly evident.

The Everglades filters rainfall and recharges the aquifers that provide clean drinking water for a third of the state. And when hurricanes spin out of the Gulf, it helps shield cities and towns against storm surges and heavy winds—a role that will become even more important as climate change causes sea levels to rise. As Rodgers, Hession, and their colleagues at the SFWMD see it, conserving the Everglades is not optional.

"It's a unique landscape," Rodgers emphasizes. "And it's not like we have another one."

The SFWMD's annual budget of more than $1.5 billion includes more than $26 million for "exotic plant control." It sounds like a lot, but it's stretched thin: the district is responsible for nearly a third of the state, including the Loxahatchee Refuge, Lake Okeechobee, and more than 2,175 miles of canals. In addition to a philosophical position against overuse of herbicides, sending crews to spray in the wetland again and again for years and years can feel like tossing cash into a hole. That's why it makes sense for the district to fund the raising and release of moths and weevils.

In balmy South Florida, lygodium reproduces year-round, and a single fertile leaflet can produce nearly 30,000 spores, each capable of traveling miles from its parent plant. But like lygo spores, arthropods also travel on the wind, and sometimes SFWMD staff come across a previously unnoticed lygodium outbreak that has already been taken care of by tiny allies.

The lygodium controls haven't yet proven as successful as those for melaleuca, but already "they're pretty much everywhere," Hession says. "I've found patches where you jump onto an island and all you see is the moths flying out of the lygodium."

There's no sign of the moths at the particular island where Hession has parked his boat, but the team of contract laborers we've come to see are doing their best without them. All Mexican nationals here on temporary visas, they typically put in ten-hour days, four days a week. It's hard work under relentless sun, in murky, hip-high water, with plenty of mosquitos and the constant possibility of encountering an alligator—or perhaps an invasive python.

There are truly few places east of the Mississippi that feel more like wilderness than the Everglades. It's not because it's so far from civilization. Nowhere in the Loxahatchee Refuge is more than six miles from either the subdivisions that run up the Atlantic coastline or the sugar plantations south of Lake Okeechobee, from which gray smoke billows over the western horizon (burning is a common practice to clear cane fields of excess vegetation before harvest). What makes this place feel wild is the fact that, without this boat, Hession, Rodgers, and I would be doomed. You can't hike out of the Everglades. You can't swim. It's an ecosystem that seems to have no use for humans. It's hard not to think that the 19th-century official who dismissed the Everglades as "suitable only for the haunt of noxious vermin or the resort of pestilent reptiles" may have had something of a point.

But even the Everglades is not exactly a primordial vestige of pure nature. In fact, humans have been making their mark on this ecosystem for millennia.

Scientists once assumed the Everglades' tree islands were all built around rocky protrusions emerging above the water level. But recent paleo-ecological research buttresses an alternate hypothesis: many tree islands may have formed over middens, prehistoric trash piles left by the humans who lived here some five thousand years ago.

"The original inhabitants of Florida moved seasonally from the west coast of the peninsula, into the Everglades, and then down into the Keys," explains Traci Ardren, an anthropology professor at the University of Miami who has conducted extensive archaeological digs in the Everglades. "They were in canoes, and along the way they were collecting resources and looking for dry ground to camp on and live for a little while."

At first, tree islands may have begun to form when vegetation snagged on outcrops of prickly limestone, attracting birds whose guano then further built up the dry land. These dry oases offered convenient stops for extended families migrating through the Everglades, who set up camps or seasonal villages sheltering perhaps twenty to thirty people. And where people settle, there will always be garbage. Excavations of these sites have turned up turtle shells and alligator bones, sharks' teeth and crab shells, broken pottery, and other human debris.

The prevailing theory, Ardren says, is that people lived on one side of the island and threw their trash on the other, then left the island to nature for the rest of the year. Over time, in this extremely fertile environment, the garbage was naturally covered by soil and vegetation, effectively expanding the island. The people would then switch sides, alternating back and forth, year after year, as they watched the dry ground grow.

"The middens are so dense—it's a lot of material," Ardren says. "I think they were deliberately augmenting the tree islands. We're just spitballing here, but we know for sure people lived on the tree islands for thousands of years. It would be human nature to become accustomed to it and assume it would continue to be available for your family, your children."

By disposing of their trash in this apparently deliberate way, the first inhabitants of the Glades were engineering a landscape that would grow into a vibrant, vital habitat for many species, eventually including non-natives like lygodium and melaleuca. In their own way, they were the forerunners of the canal-dredgers and swamp drainers who showed up millennia later to create new homes for the millions of people who would flock to Florida in the 20th and 21st centuries. The consequences then were more dire, but the intentions, perhaps, the same.

The original Florida tribes died out after the Spanish arrived in the 16th century, bringing with them European diseases. The Miccosukee and Seminole who fled Georgia and resisted displacement at the hands of the US government couldn't have known that these ancient people had helped to create the dry ground where they found refuge—or that the impact of humanity on the environment stretches even to the places that feel most like wilderness.

6

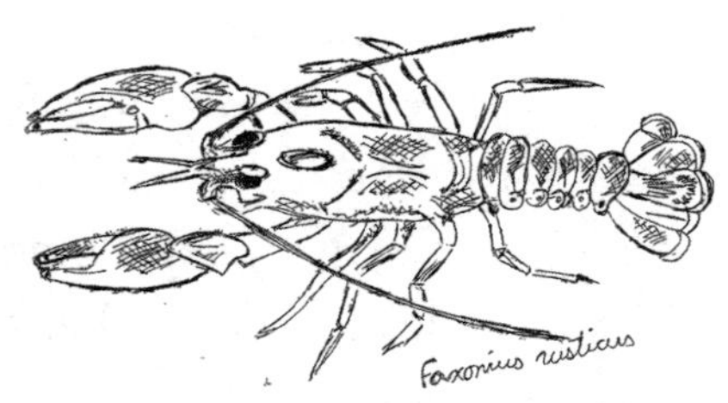

BORDER CROSSINGS

A hiker stumbling across a crayfish for the first time in one of Ontario's glacial lakes or frigid creeks might be forgiven for thinking the aquatic arthropod seemed a little out of place. With its bright eyes, fanning tail, nervous antennae, and fierce claws, a crayfish looks like a miniature lobster, perhaps more suited to skittering along the seabed than enduring the extremes of Canada's four seasons.

On the other hand, while crayfish are actually quite common worldwide, anyone who isn't keeping a careful eye out for these well-camouflaged, elusive crustaceans would be unlikely to find one at all. There's a certain skill involved.

"I had to sort of learn on the job, because nobody even knew where these crayfish were and what they do and which species they are and all that," says Radu Guiaşu, a biologist at Toronto's York University. His 1997 doctoral dissertation broke ground with a survey of the behavior, distribution, and status of Ontario's ten native crayfish species, and today he is among Canada's foremost experts on the armored invertebrates.

In the remote natural areas where he conducts fieldwork, Guiaşu's striking salt-and-pepper mane and matching bushy beard might let him pass as a modern-day version of the 19th-century mountain men who made their livings in the Canadian fur trade. Like them, he's also an experienced trapper, although his traps are baited with a modern lure: fish-flavored cat food. ("Chicken-flavored cat food doesn't work as well," Guiaşu deadpans, "probably because not a lot of chickens fall in the water.")

His preferred method of collecting crayfish, however, requires nothing but a sharp eye, a net, and a little finesse: "Their biggest muscle is in the tail, so they tend to backflip away from you, and if you scare them with a stick or something, they flip into the net."

Back at the lab, Guiaşu keeps wild-caught crayfish in tanks of fresh water, where he and his students study their behavior and sensory capabilities. Crayfish might not be the most charismatic species, but years of close observation have persuaded Guiaşu that they rank among nature's more fascinating creatures.

"Most of them are loners, and they're very aggressive," Guiaşu says. "They tend to fight, but when a crayfish loses a fight it signals the loss by tapping the head of the opponent with the antennae twice. It's a signal to say, 'Okay, let's stop. I lost. I'm going to go away.' It's not always that simple, but after a while, you get to decipher their language, and their language is mostly chemical and tactile."

Like most arthropods, crayfish don't have particularly sharp vision. But with their antennae and a smaller pair of sense organs just above, called antennules, they can detect a vast range of chemicals, including intraspecific pheromones that indicate whether other crayfish are interested in mating—or fighting. This ability, which scientists believe actually improves as they grow older, likely provides crayfish with the olfactory equivalent of a technicolor experience.

"The antennules get longer with age, with every molt, and they keep adding sensory structures," Guiaşu says. "When we walk through water, we can maybe tell if it's cold, but that's about it. Crayfish can actually smell the water."

Their exquisite sensitivity makes crayfish especially interesting to ecologists, who see their presence as indicative of habitat quality. And crayfish, which eat mostly detritus and are in turn eaten by fish, birds, snakes, and mammals, are a keystone organism in freshwater food webs, required for the proper function of entire ecosystems.

In Ontario, however, not all crayfish are considered equal.

Muddy Waters

Among its North American relatives, the rusty crayfish (*Faxonius rusticus*, formerly assigned to the genus *Orconectes*) doesn't immediately stand out. About the length of a man's middle finger, it's best identified—albeit only when dealing with adults—by the rust-colored patches on either side of its shell, just above the tail. But this unassuming crustacean faces an outsized rap sheet.

Canadian authorities have long assumed the rusty crayfish to be an American import, brought across the border by anglers for use as bait. It's a slightly surprising origin story, because among the charges brought against *F. rusticus* is inedibility, at least where sport fish are concerned.

"At a large adult size of 10 centimeters, it is too big to be eaten by most native fish," explains a recent blog post published by the advocacy group Trout Unlimited Canada. "Additionally, when faced with a potential predator, Rusty Crayfish will take an attack stance with its big, reddish claws raised above its head—quite the intimidating display for such a little critter! This often scares off any would-be predators, allowing them to colonize new areas with minimal competition."

A stubborn refusal to be eaten is not the only complaint lodged against *F. rusticus*. "Rusty crayfish can feed at a rate of twice that of a native crayfish species," warns the Nottawasaga Valley Conservation Authority, an Ontario land management agency. "This decrease in food availability can cause local extinctions of native crayfish species! As well, a female can hold up to 200 fertilized eggs, allowing a single rusty crayfish to quickly establish a population in a new area."

Aggressive, voracious, and hyper-fertile, rusty crayfish check all the boxes required for a problematic invasive species. And Premek Hamr, an independent scholar based in Ontario's rural Kawartha Lakes region, has been tracking their spread across the province with increasing dismay.

"In the mid-eighties, they were just starting to take over southern Ontario," says Hamr, who works out of a cozy home office lined with bubbling fish tanks and plastered with crayfish posters. "Certain waterways were already infested with them, and basically not a lot was done."

Known online as "Dr. Crayfish," Hamr has a deep affection for Canada's native crayfish, which he has been studying for decades. He speaks lovingly of creatures like the Appalachian brook crayfish (*Cambarus bartonii*) with its "Porsche"-like streamlining, and the tough-looking big water crayfish (*C. robustus*), Canada's largest.

But the rusty is another matter. Hamr worries that it's taking over.

"I did some surveys in the nineties and early 2000s, and there were more and more of them," Hamr says. "There were so many!

One of the highest counts I did was just over a hundred per square meter. *Per square meter*!"

Hamr traces his crustacean fascination to an early memory of chasing crabs on the beach in the West African nation of Ghana, where his father was posted as a diplomat from Czechoslovakia in the late 1950s. It was an exciting, peripatetic boyhood that came to an abrupt end in 1968, when the liberal reforms of the "Prague Spring" prompted the USSR to crack down on its recalcitrant satellite state. Posted in Turkey by then, the Hamrs defected to Canada, where they were granted political asylum.

Hamr, who was thirteen when his family settled permanently in Montreal, went on to study environmental science in college, where a chance encounter during a field trip to a local pond hooked him on crayfish. In 1985, he left Canada to pursue a PhD in Australia, lured by the Tasmanian giant crayfish (*Astacopsis gouldi*), which, at up to thirteen pounds—the size of a Jack Russell terrier—is the largest freshwater invertebrate in the world. In Tasmania, the government used Hamr's research to support official protection for the unique creature, whose slow rates of growth and reproduction make it especially vulnerable to habitat loss and overhunting.

"I was the first person to study the giant crayfish, so I gained fame that way with the crayfish community," Hamr says. "But when I came back to Canada in 1990 and tried to get a job in my field, nobody was interested in crayfish."

Unable to secure a tenure-track position, Hamr cobbled together gigs as a government contractor and university lecturer before spending two decades teaching science at Upper Canada College, a top boys' prep school in Toronto. Still crazy about crayfish, he conducted his own research in Ontario's lakes and streams each summer, recruiting student helpers and occasionally contributing to scholarly publications. Then, around the time he retired from teaching in 2020, "crayfish hit the big time," Hamr says, "because of all these invasions."

The rusty crayfish may be a marquee invader, but other non-native species that have Canadian authorities on high alert include the marbled crayfish (*Procambarus virginalis*), an aquarium-bred oddity that reproduces by cloning itself and which recently prompted a frantic bout of pond-draining in Burlington, Ontario. For Hamr, these interloping invertebrates have meant steady consulting work for government agencies and environmental nonprofits, which are working to track, prevent, and spread the word about invasive species. And in 2024, after decades on the fringes of academia, he was appointed an adjunct professor at Trent University, in Petersborough, where he now supervises student research.

Hamr's dedication to crayfish may finally have been vindicated, but the late-life career revival is bittersweet. There's truly no chance that the rusty crayfish can be expelled from Canada, Hamr acknowledges. All he can do is work with provincial authorities and conservation groups to tamp down new outbreaks and to educate the public about the danger of moving crayfish from one waterway to another—a danger recently recognized in law, as several of Canada's provinces have prohibited anyone from carrying crayfish overland (unless, like Hamr, they have a research permit).

"There was a big expansion of rusty crayfish before they banned the sale," Hamr says. "Once they banned the sale in bait shops, that slowed it down a little bit. But now they've spread into Quebec, and there are big concerns."

Cold comfort though it may be for Hamr and other native crayfish advocates, the official position across Canada is now one of fierce opposition to the rusty menace. But there remain a few complications. For one thing, it's not entirely clear where the rusty crayfish does belong. While Hamr points to maps that show it only recently expanding across Ontario and into Quebec, his fellow crayfish expert Guiaşu notes that sources disagree about the true origins of *F. rusticus*. Indeed, Guiaşu argues that when examined closely the boundaries of its purported range can seem somewhat arbitrary.

"The rusty crayfish is considered native to parts of Michigan, but not all of Michigan," Guiaşu says. "And it's considered native to parts of Ohio—but not all of Ohio. And you drive four hours to Ontario, all of a sudden it's a bad, introduced crayfish."

The rusty crayfish was first described by a Smithsonian Institution biologist named Charles Girard in 1852 from a preserved specimen that had been collected in Cincinnati and sent to his Washington, DC, laboratory. Since then, the rusty has been considered a native species in the Ohio River watershed, where it is established and abundant. But is that its only legitimate home?

The US Geological Survey (USGS), which tracks non-native aquatic species, offers the closest thing to an official map of the rusty's native range, but its staggering precision raises more questions than it answers. The map includes much of Ohio, stretching all the way to the shores of Lake Erie—although *F. rusticus* is deemed invasive in the lake itself, despite having been recorded there as early as 1897. The rusty is acceptable in the Maumee River, which meets Lake Erie at Toledo, but unwanted in the Ottawa River, which also originates in Ohio and drains into the same bay less than two miles away, just across the Michigan line. At the opposite end of the rusty's range, in Kentucky, the USGS declares it native in creeks that pass just two miles north of man-made Lake Cumberland, but not in the nearby network of waterways that feed the reservoir itself (the map records as invasive an *F. rusticus* specimen collected there in 1982).

Meanwhile, the USGS isn't necessarily in agreement with other sources. In 2021 Guiaşu and a student conducted a literature review, looking at 430 scientific publications from as far back as 1852. While a plurality of references did place the rusty's native range in Ohio, others also counted *F. rusticus* as native in territories as far-flung as Mississippi and Minnesota. Ten studies the pair turned up actually considered rusties native to Ontario.

In that province, the rusty's historical distribution has been traced largely through specimens in the collection of the Royal Ontario Museum, but Guiaşu has found these records to be frustratingly patchy. Although the museum's crayfish database is the largest and most comprehensive in the province, the entirety of its rusty specimens were gathered over just twenty-five days in thirteen different years, with the majority dating to the 1960s. Nearly half of the collection dates to just one trip, in 1964—the one that supplied most of the research for the museum's 1968 *Handbook of the Crayfishes of Ontario*.

The first publication to identify the rusty crayfish within the province, the *Handbook* was also the first to determine that it was "clearly an introduced species," although evidence for introduction was circumstantial: the authors connected "large numbers" of rusty crayfish in an area of southern Ontario with the fact that "most of the summer vacationers" there at that time hailed from the rusty's alleged home state of Ohio. But did the rusty crayfish first appear in Ontario in the 1960s? Or is that simply the first time scientists looked for it?

Like other aquatic animals collected for scientific purposes, crayfish specimens are preserved in jars of alcohol, a preparation that eventually drains them of their color. Briefly labeled with species and location information, these spectral arthropods wind up lining shelves in florescent-lit storage facilities where, for the most part, they are left alone. But over the years,

Guiaşu has examined thousands, including the entire collections at the Royal Ontario Museum and at the Canadian Museum of Nature in Ottawa. ("I do have a life!" he says with a laugh.) Even such an exhaustive search, however, turns up few crayfish from before 1900. That makes it hard to say which species were where before cities, industry, and agriculture began altering Ontario's environment—let alone before European settlers arrived.

"Virtually no one was looking for crayfish in the early days," Guiaşu says. "So, we just don't know."

What scientists do know is that crayfish, like many other modern plants and animals, have been expanding north since this continent's glaciers began retreating at the end of the last ice age, about twelve thousand years ago. Crayfish remain most numerous and diverse in the southern United States, which boasts 330 different species (including two only identified in 2024), but they are resourceful creatures that can easily move within connected waterways, sometimes even traveling short distances overland. The absence of a certain species from a particular spot today, Guiaşu argues, doesn't mean it won't ever get there on its own. And if it does, should we consider it a foreigner?

For Guiaşu, the transformation of the rusty crayfish into a public enemy has been a persistent source of frustration, informing his 2016 book *Non-native Species and Their Role in the Environment.* Now squarely in the camp of the invasion biology apostates, he sees his position as more in keeping with a true appreciation for nature, with all its mysteries. It's a feeling he forged as a child, exploring the mountains outside Bucharest, Romania, in the waning days of the communist era.

"I really miss mountains," he says, noting that Ontario, while ruggedly beautiful in its own way, lacks his homeland's dramatic topography. "In Romania, no big city was very far from the mountains. You started in the plains, then hills, then bigger hills, and then within an hour and a half, on a very slow train, you were in the mountains. It was like a geography lesson. And when I was walking through the forest, nobody was pointing out, 'This flower doesn't belong,' or 'This bird is from elsewhere.'"

Despite Guiaşu's bucolic memories, the Romania of his childhood was suffering under the brutal regime of socialist dictator Nicolae Ceaușescu. Like Premek Hamr's family, Guiaşu's parents, both academics, eventually emigrated with their son to Canada, where the teenage refugee encountered plants and animals that reminded him of home.

But something about them had changed.

"Species that I grew up with in Europe, all of a sudden they were demonized in North America," Guiaşu says. "And to me, they're the same species. Dandelions, house sparrows, things like that. All of a sudden they turned bad because they crossed the ocean."

It's a phenomenon that has in turn bemused, alienated, and disturbed him.

As in the United States, Canadian conservation organizations often focus the time and energy of staff and volunteers on campaigns to eliminate one invasive species or another. Given his reputation as a crayfish enthusiast, Guiaşu was once invited to participate in an event the organizers billed as a celebration of biodiversity. The day's main activities were a garlic mustard pull and a hunt for rusty crayfish.

"I said, 'Look, I'm sorry, but killing plants and animals doesn't seem to me like the best way of celebrating biodiversity,'" Guiaşu says. "I think it sort of poisons people's attitudes toward nature, and it seems like there are better ways to spend the limited resources we have for conservation. The rusty crayfish is basically just another crayfish. The ecology of the rusty crayfish is the same as any other local crayfish species that I can think of. And I studied these for decades."

Such equanimity remains unfathomable to Hamr, who sees the rusty invasion as a slow-rolling ecological catastrophe.

"People say 'Well, crayfish is crayfish. Why worry about it when you have lots of crayfish in the stream?'" he says. "Okay, so if you're a birder, would you be happy to replace all the birds with just pigeons because they're birds? Or, I should really say gulls, right? Because the rusty crayfish are more like gulls. They're aggressive, and they spread, and they're in huge numbers."

What it comes down to, Hamr fears, is the loss of distinction, a sort of gray goo spreading through the world as gradually every crayfish in the ponds and brooks across Ontario starts to look exactly the same.

Swamp Creatures

Hamr's exquisite sensitivity to crayfish diversity is likely lost on the non-expert, for whom closely related crayfish can be difficult to tell apart. On the other hand, the rare white morph of the northern clearwater (*Faxonius propinquus*), which has only ever been found in the crystal waters of Ontario's Lake Simcoe, is unmistakable: ghostly pale except for a dark patch on the top of its tail, as if someone had dipped it claws-first into a bucket of bleach. But *F. propinquus*, including its basic brown morph, is declining in Lake Simcoe. The suspect, once again, is the rusty crayfish, a congeneric that has been noted to interbreed with this Canadian native.

While the hybrids resulting from such couplings are fertile, as well as larger and more vigorous than either parent, researchers studying crayfish captured in one northern Wisconsin lake found that, when backcrossed with either parent species, they produced comparatively fewer offspring. And these, unlike first-generation hybrids, were actually *less* likely to survive. The researchers linked the phenomenon, as observed in their lab, with a reduction in *F. propinquus* in the Wisconsin lake and reasoned that it also occurred in other places where the two species come into contact.

Hybridization has long been a concern of invasion biology, which warns that interbreeding with introduced species can have a negative impact on native populations. In one scenario, hybrids may turn out to be less fit than their parents, essentially wasting the reproductive efforts of a more vulnerable native species to the relative advantage of the invader. Another possible mechanism for so-called extinction by breeding is introgression, also known as "genetic swamping," the gradual replacement of a rare species' genes with those of a more common or widespread relative through repeated backcrossing.

Introgression has been reported in Scottish wildcats (a population of the European wildcat *Felis silvestris*), which hybridize with feral housecats (*F. catus*), as well as Spain's endangered white-headed ducks (*Oxyura leucocephala*), whose tendency to interbreed with introduced North American ruddy ducks (*Oxyura jamaicensis*) has united Europe in a campaign to eradicate the latter. But perhaps the best-known case comes out of the deep history of our own species. Neanderthals (*Homo neanderthalensis*) lived in Eurasia for hundreds of thousands of years, until they all but disappeared from the fossil record around forty thousand years ago, not too long after a major dispersal of modern humans (*H. sapiens*) out of Africa. Competition and conflict between our species likely played a role, but recent studies of the human genome have found that most of us today carry a small percentage of Neanderthal DNA in our own cells. This means that our two species interbred, perhaps leading to the gradual, and almost total, replacement of the Neanderthal genome with our own.

In *Homo*, at least, hybridization seems to have reduced biodiversity, but in general it's hard to predict where interbreeding between distinct lineages will lead, especially over the very long term. Hybridization can be a destructive force but also a creative one—and, indeed, some famous biodiversity hotspots may owe their wealth to it.

East Africa's Great Lakes are home to a group of fish that has long captivated both scientists and aquarium hobbyists. Found in tropical regions around the world, cichlids are among the largest vertebrate families, representing at least 1,650 different species—and possibly twice that amount, as new ones seem to turn up every year. But nowhere are cichlids more diverse than in Lake Malawi, which boasts at least 850. They include hot-tempered black-and-yellow "bumblebees" (*Pseudotropheus crabro*); parti-colored "peacocks" (genus *Aulonocara*), which rival the denizens of any tropical reef; and the mysterious predator *Pallidochromis tokoloshe*, which dives to depths of nearly five hundred feet and gets its specific epithet from Zulu mythology's evil water spirit Tokoloshe.

The astounding variety of Malawian cichlids arose through a process called adaptive radiation after a common founder population arrived some eight hundred thousand years ago. Finding themselves in a lake with many different ecological niches—chilly depths, rocky coves, sandy flats—the ancestors of today's cichlids adapted to better fit the diversity of their environment. But while that's textbook evolution by natural selection, most groups of organisms do not diverge so rapidly. Malawi's first cichlids must have had some advantage hidden in their genome—and that, researchers believe, may be a history of hybridization.

In 2020, an international team analyzed the genomes of Malawian cichlids, finding evidence that their rapid diversification was fueled by hybridization between two distinct lineages that had originally split between three and four million years ago. Admixture introduced novelty into their genomes when these lineages reconvened in the lake, providing what the researchers called "genetic raw material" upon which selection acted to induce rapid speciation.

"A clear message," the researchers wrote in the journal *Molecular Biology and Evolution*, "is that hybridization between related species is far more common than previously thought, leading to fundamentally reticulate patterns of evolution, where species relationships are represented by networks rather than trees."

Genomic research is revealing more and more such patterns of divergence and recombination. And in the sped-up Anthropocene era, as humans export organisms around the world, it's a process that we can now watch in real time.

In 2008, when Mario Vallejo-Marin was a newly minted PhD, the Mexican-born scientist moved to Scotland to join the faculty at the University of Stirling. The medieval city, its skyline dominated by a 12th-century castle, is perhaps an unlikely spot to encounter monkey flowers, bright and cheerful blooms named for their putative resemblance to a monkey's face. None of the several species in the genus *Mimulus* is, after all, native to Europe. But like many pretty plants from exotic locales (in this case, North America, Australia, Africa, and Asia), monkey flowers caught the eye of Victorian gardening enthusiasts, who eagerly imported them to the United Kingdom.

"You find them everywhere," says Vallejo-Marin, now a professor of ecological botany at Uppsala University in Sweden. "Not only do you have North American species, but also a South American close relative. In the Americas they never meet—they don't have a chance, they're separated by thousands of kilometers. But in my neighborhood, I could see them growing side by side."

M. guttatus, from North America, and the South American *M. luteus* are also separated by about four million years of evolution, but in Scotland they frequently hybridize. The hybrids generally wind up with an extra set of chromosomes, making them "triploid," but they are robust and vigorous, with showy yellow flowers, and while they produce no seeds they can spread clonally. But in 2011, on a hiking trip with his family, Vallejo-Marin collected an interesting sample from a streambank in Scotland's Lowther Hills. Clearly a hybrid, this monkey flower was nevertheless producing seeds. Back at his lab, these seeds grew into new, healthy plants, which in turn produced their own seeds.

It turned out the plant had gained yet another set of chromosomes, for a total of four, and with that had come the genetic stability necessary for sexual propagation. Reproductively isolated from its *Mimulus* relatives thanks to the bonus sets of chromosomes, this was, in fact, a new species—and not just new to science. Vallejo-Marin calculated that it was no more than 140 years old, making this monkey flower one of the newest species on the planet. He named it *Mimulus peregrinus*; the specific name is Latin for "foreigner."

As human activity facilitates more reunions between related species—so-called sister taxa—biologists are realizing that many more can interbreed than previously believed. But *M. peregrinus* offers further evidence that remixing genomes can actually spur speciation. It's become accepted science that hybridization is not rare, Vallejo-Marin says. Now, scientists are trying to figure out how important it is to the process of evolution.

"I think it's important, at least in certain groups, as an engine for creating or generating genetic variation that then can be acted upon by selection," Vallejo-Marin says. "When you get a pool of individuals with very different genetic constitutions and then you bring them together and the outcome is some highly variable population, well, variation is kind of like the raw material for evolution. Then you're going to accelerate evolutionary change, explore different parts of the adaptive landscape that you couldn't explore before."

The Malawian cichlids are a prime example of hybridization spurring evolution, he notes, but they're only one of several interesting cases—including South American butterflies and even Darwin's famous Galapagos finches—that recent gene-sequencing research has revealed. It's a phenomenon that would seem to disprove invasion biologists' concern that more frequent hybridization will lead to a loss of biodiversity, but both can be true, Vallejo-Marin says.

"On the surface it sounds contradictory, but I think what makes the key point here is the timescale at which these two things happen," he explains. "In order for hybridization to function as a creative force, you are thinking about relatively long timescales. It might be fast in evolutionary terms, but still we're talking about tens of thousands of years. The eroding effect of hybridization on biodiversity is the one that occurs very rapidly. What you have nowadays is that you erode these physical barriers that keep species apart in a matter of years or decades. And there is not sufficient time for evolution to recreate this variation. At least, not in the short term."

Hard Lines

M. peregrinus bears a striking flower, canary yellow with bold red spots running down its throat. But the most interesting thing about it is that it offers science the chance to witness evolution on a human timescale. Normally, after all, nature has as little regard for our short lives as it does for the categories we impose upon it.

"Species," for example, becomes trickier to map the longer you look at it. By some counts, there are dozens of possible meanings for the term, but the classic one, as still generally taught in biology classes, is the so-called Biological Species Concept proposed by evolutionary biologist Ernst Mayr in 1942. It hinges on reproductive isolation—that is, under this definition, a species is a set of organisms that only breed successfully with one another, whether because their unique physiology, genetics, or behavior prevent productive couplings with outsiders or because they are separated from closely related groups by longstanding and insurmountable geographical barriers. But there are obvious exceptions to this elegant rule.

Organisms that reproduce asexually, including bacteria, certain fungi and plants, as well as unusual animals like the self-cloning marbled crayfish, clearly aren't covered by Mayr's concept. Taxonomists have classified these into species based on observed characteristics or, more recently, genetic similarity—although some question how "species" can be a meaningful category at all for prokaryotes (single-celled organisms without nuclei, such as bacteria), given the frequency of horizontal gene transfer between even distantly related strains. Meanwhile, extinct organisms, which may or may not have once interbred, are divided into species based only on the morphology of their fossils, using criteria that even paleontologists admit can seem arbitrary.

In the 21st century, new gene-sequencing technology has introduced greater nuance into taxonomy, as researchers identify many so-called cryptic species that are distinguishable from one another only at the level of DNA. The recent flood of such publications has led some to estimate that as many as 30 percent of currently designated species may, by dint of genetic distance, actually be more than one. Genomic analysis has convinced scientists, for example, that African elephants actually consist of two separate species, the savanna elephant (*Loxodonta africana*) and the smaller forest elephant (*L. cyclotis*), which diverged genetically several million years ago.

The matter has little relevance to the way these elephants live their lives today—indeed, research indicates they blithely interbreed where their ranges overlap. For humans, though, categories and divisions represent more than just our attempts to make sense of the world. They are also political tools, and they can be very useful for determining priorities and allocating resources. Before 2021, for example, the International Union for Conservation of Nature listed African elephants as a single species—one that was "vulnerable," a classification indicating only moderate urgency. Upon dividing them into two, the IUCN recognized the situation as more dire: savanna elephants became "endangered" and forest elephants "critically" so. The move made international headlines, and conservation groups hoped it would draw more attention and resources to elephants' plight.

Sometimes, genomic sequencing seems to move speciation in reverse. A 2025 paper published in the journal *Current Biology*, for instance, declared that the snail darter (*Percina tanasi*), a small fish discovered in Tennessee in 1973 and soon afterward declared an endangered species under the newly enacted United States Endangered Species Act, is actually just a population of the more common stargazing darter (*Percina uranidea*). By then, however, the fish had served its purpose. Conservationists used the alleged snail darter's alleged plight to block the construction of a dam near Knoxville, an impasse which it would take an act of Congress to overcome (the dam was, in fact, constructed in 1979).

"I feel it was the first and probably the most famous example of what I would call the 'conservation species concept,' where people are going to decide a species should be distinct because it will have a downstream conservation implication," Thomas Near, a Yale ichthyologist and author on the *Current Biology* paper, told *The New York Times*.

If the borders humans draw on nature can have political consequences, drawing political borders can also have consequences for the natural world.

The Rio Grande, which makes up more than 1,200 miles of the border between the United States and Mexico, is among the most politically charged waterways in the world. Unlike the quieter frontier with Canada, which the rusty crayfish may—or may not—have slipped through unnoticed, this border sees a flurry of activity. Encounter records released by US Customs and Border Protection, or CBP, show the number of human migrants crossing this border without permission roughly quintupled between 2017 and 2024. The situation, which recalled previous "border crises" dating back decades, was among the top issues in the 2024 presidential election.

As in the 1970s, when Central Americans fled a series of civil wars, people hoping to cross the border today are driven north by forces beyond the control of the individuals who make a difficult choice to leave their homes. They include poverty and crime, but also climate change, which is worsening natural disasters and altering the weather patterns upon which Latin American farmers depend. And this stream of human migration in turn contributes to an environmental crisis along the border, with both migrants and law enforcement leaving their marks on the natural world.

Humans, meanwhile, are not the only ones crossing the border—nor is *Homo sapiens* CBP's sole species of interest.

In Texas, the border region has also become a major battleground in the war on invasive species, a fight that CBP has joined as agents target giant reed (*Arundo donax*, known locally as carrizo cane), which is native to the Middle East and parts of Asia. The plant, imported to the Americas as building material perhaps as long ago as the 1600s, is accused of overusing water, displacing native riparian habitat and sheltering the ticks that transmit deadly cattle fever, a disease endemic to Mexico that is the subject of a draconian quarantine north of the border. The twenty-foot cane also provides excellent cover for migrants and drug smugglers along the muddy banks of the Rio Grande.

Comparing the ways in which we talk about human border crossing and the spread of invasive species can at times feel so obvious as to be tiresome, but at the border between the United States and Mexico it becomes clear that, in the 21st century, the two issues are not merely parallel but intertwined. And the connections go beyond the border itself.

On September 11, 2001, a little more than two years after President Clinton signed Executive Order 13112, aiming government resources at undesirable plants and animals, Islamist terrorists attacked the United States, killing nearly three thousand people and prompting the US government to declare an all-out "War on Terror." This spiraling global conflict would soon overlap with both the war on invasive species and the fight against illegal immigration, as authorities grew increasingly concerned not only that terrorists could sneak across America's borders but that they might bring along non-native plants, animals, or pathogens for use as weapons.

The introduction to the 2013 interdisciplinary volume *Biosecurity: The Socio-Politics of Invasive Species and Infectious Diseases* summed up the mood, declaring, "[B]iosecurity and associated issues, such as bioinvasion and nativism, are emblematic issues for the twenty-first century." The idea that lifeforms such as invasive plants, agricultural pests, and deadly diseases can and should be prevented from crossing human-drawn borders, the editors noted, touched on hallmark 21st-century concerns, including about localism versus globalization and security in an uncertain future. (Daniel Simberloff himself contributed a chapter titled "A world in peril? The case for containment.")

The novel term "biosecurity" also conformed to the new, tense, and somewhat paranoiac lexicon that emerged in the post-9/11 years: terms like "homeland," "counterinsurgency," and "weapon of mass destruction," which trickled into the vocabulary of journalists, officials, and laypeople much as the jargon of the COVID-19 pandemic—"super-spreader event," "social distancing," "shelter-in-place"—would two decades later.

"Border security is national security—there is no difference—and the crisis on our southwest border puts our national security at risk," US Border Patrol Chief Carla Provost told a Congressional committee in 2019, deploying a very 21st-century style that was eerily echoed the next year by an op-ed in the journal *Science*. Penned, again, by Simberloff, it bemoaned the Trump administration's decrease in federal spending for invasive species control, alleging that the move put "national security at risk by lowering barriers to these devastating invaders."

The invaders to which Simberloff referred were, of course, not humans bent on attacking America, but plants and animals that he and his coauthors asserted could "cause harm" to anything from our "economic, food, water, and infrastructure security to human and environmental health to military readiness." Heeding such warnings, President Joe Biden would eventually restore the funding, a move his US Fish and Wildlife Service Director

Martha Williams touted as "an opportunity for us to invest in eradication efforts that protect our lands and waters."

To return, briefly, to Canada: when the rusty crayfish takes its "intimidating . . . attack stance" in the clear, cold waters of an Ontario stream hundreds of miles from the fraught Rio Grande, applying a biosecurity framework to invasive species may seem like overkill.

But in another commonwealth nation halfway around the globe, a foreign crayfish might be cause for national alarm. There, biosecurity is an all-encompassing worldview, one with implications for an entire society, and perhaps the world.

7

PREDATOR FREE

Serving as official naturalist on Captain James Cook's first around-the-world voyage (1768–71), Sir Joseph Banks was the first Westerner to describe kangaroos, deriving the name from an Aboriginal word that was likely closer to "gangaru." In addition to these marquee members of the *Macropodidae* family, Banks would sail back to England with records of other peculiar Australian fauna, including the dingo and the quoll, as well as another group of animals that he labeled "possums" (family *Phalangeridae*).

Like any learned Englishman of his era, Banks was aware of North America's opossum, which was by that point a famous oddity. So when he and his scientific staff came across a nocturnal, arboreal marsupial of roughly the same size, with a longish tail and a pointed face, they decided it must be a similar sort of beast. It later turned out that American opossums are only very distantly related, but the name stuck, inaugurating centuries of confusion.

Not in New Zealand, though. Here, people know what possums are—and they know what to do with them.

"Best method for killing possums is to head-shoot them," advises the New Zealand Deer Stalkers Association, which recommends spotlighting the beasts at night with a powerful LED flashlight. "You generally have two bright red eyes to aim at."

Once you have a small pile of dead possums, you can get to work plucking them—although this has to be done quickly, because the task becomes more difficult as the possums' bodies cool. Used in knitwear, possum fur is reputedly warmer and lighter than sheep's wool, and a thriving industry has spun up around it in recent years (according to the New Zealand Institute of Economic Research, possum products generate up to NZ$150 million in annual retail sales). Possum hides are also marketable, and as long as the animal hasn't been poisoned—another accepted method of dispatch—the meat has potential as an export to Asia. Best of all, possum products are "green and ethically sound," boasts the New Zealand Fur Council, which promises, "Possum harvesting can greatly improve flora and fauna growth and preservation at zero cost to taxpayers."

As New Zealand's attitude toward these creatures makes evident, possums are not native to this southern archipelago. Indeed, no terrestrial mammals are, except for two small bat species. When the Polynesian ancestors of today's Māori arrived in the land they would call "Aotearoa" ("Long White Cloud") in the second half of the 13th century CE, New Zealand had been geographically isolated for eighty million years, and large, flightless birds known as the moa (order *Dinornithiformes*) dominated the countryside, preyed on only by Haast's eagle (*Hieraaetus moorei*), a massive raptor that topped out at thirty pounds.

Māori hunters drove these giants to extinction long before European explorers set eyes on Aotearoa, but New Zealand's first people reshaped the landscape in other ways, too. With fire and axes, they carved the native forest into grasslands, which were more convenient for hunting and farming. Verdant meadows greeted British settlers when they began arriving in large numbers after 1840, and these newcomers set to work building an economy based upon livestock farming. Soon, sheep and cattle would range across the land by the millions.

Around the middle of the 19th century, a few colonists had the bright idea to import the common brushtail possum (*Trichosurus vulpecula*) from Australia, in hopes that the fuzzy creatures could serve as the foundation of a fur industry. Their venture was ultimately not successful, but once possums had traveled the nearly one thousand miles across the Tasman Sea it was impractical to send them back. And anyway, they took to their adopted home quickly, spreading across both main islands—gradually at first, then all of a sudden.

By the 1970s, possums could be found in nearly every corner of the country.

By the end of the 20th century, they were "public enemy no. 1."

In 1999, the New Zealand Tourism Board debuted its official slogan, "100% Pure New Zealand." Originally the centerpiece of an advertising campaign meant to entice visitors with images of craggy peaks and turquoise bays, the slogan has endured for a quarter century because it also reflects Kiwis' idea of themselves and their nation. And if New Zealand is the planet's last vestige of unspoiled natural beauty, invasive species go against the brand.

"We relate our national identity to our native wildlife," Nicola Toki, the chief executive of the nonprofit Forest & Bird and a public face of New Zealand conservation, told *The Atlantic* magazine, which reported on the matter for a US audience in 2013. "And everybody hates possums."

Indeed, possums are a top target of the national campaign against invasive mammals known as Predator Free 2050. There are an estimated thirty million of them across Aotearoa, six times the human population, and they face official blame for stripping native vegetation, crowding birds out of potential nesting sites, and spreading bovine tuberculosis, which threatens the economically important dairy industry. While possums are herbivorous, with digestive systems geared toward fibrous leaves, video evidence shows that they occasionally eat bird eggs and even chicks as well.

Wiping out possums—and other non-native mammals, including stoats, rats, and ferrets—will be difficult, but public opinion in New Zealand runs in favor of the effort. To adopt a trendy term from international relations, the country has taken a "whole-of-society" approach to the issue, as national and local government agencies, nonprofits, businesses, and regular citizens across the archipelago do their part to return Aotearoa to its original, mammal-less state (well, almost: humans, pets, and livestock are, for now at least, welcome to stay).

It's an effort that has united New Zealanders from every generation and walk of life, and if successful it could prove an inspiration for other environmental campaigns around the world. But in a country whose inhabitants are so attached to their native species that they proudly bear the demonym "Kiwi," the embrace of biosecurity as a patriotic goal also raises thorny questions about nationalism, human psychology, and the true cost of turning back time.

Traplines

Before she immigrated to New Zealand to 2010, Annalily van den Broeke had very different feelings toward small mammals.

"I've always loved animals," she says, in an accent that's now halfway between crisp Dutch and reedy Kiwi. "And as a volunteer in the Netherlands, I worked at what they probably here would call a pest animal shelter." Van den Broeke has the sun-kissed face of someone who spends a lot of time outdoors, and her eyes crinkle as she laughs, thinking of the abandoned rabbits, rats, mice, chinchillas, and guinea pigs she lovingly cared for until they found forever homes. "Then you come here and you really have to do a mind switch."

It helped, she says, to think of pests and pets as categorically different.

"'Rabbits' are the ones in the field out there, a pest animal," she explains. "'Bunnies' are the ones in a cage, in a house, that never go outside. They're desexed, they're vaccinated. Same with rats. Pet rats are the same species, but they're different from the ones that you have to get rid of."

Van den Broeke and her husband were drawn to New Zealand's natural beauty, and one of the first things they did after settling in a suburb of Auckland, the country's largest city, was to sign up as volunteers with a local conservation project called Ark in the Park. It would be a way to spend time outdoors, meet people with similar interests, and, most of all, help care for the beautiful landscape and charming creatures that had lured them to the other side of the globe in the first place. But taking care of nature in Aotearoa, the softhearted vegetarian van den Broeke soon discovered, nearly always means killing animals.

It was an unpleasant prospect. She would have to ease her way into it.

"The thing that you do first as a newbie is putting out poison for rats," she says. "It's the first step, and it is something that a lot of people are used to anyway, because they're also a nuisance in your house. What we use are blue pellets. They come in a plastic bag, and they smell nicely of cinnamon. You go into the forest, and there are certain plastic containers that you refill with this bag. You don't handle the poison. You don't see dead animals. You just have this beautiful walk through the forest. And a lot of people stop there and just do that."

The next step is trapping, van den Broeke says, and that's a bigger step.

"Once you trap something, you have to get it out of the trap," she explains, "so you *are* handling dead animals."

The more time van den Broeke spent in the forest, however, the more she saw the importance of this admittedly gruesome work.

The earliest Māori settlers deliberately imported Polynesian rats (*Rattus exulans*) as a food source. Known locally as "kiore," the high-protein rodents are considered a delicacy, but their spread through New Zealand has been associated with declines in frog, insect, seabird, and reptile populations, as well as changes in vegetation. Later, black (*R. rattus*) and Norway (*R. norvegicus*) rats stowed away with European explorers and—along with house mice (*Mus musculus*)—eventually displaced their Polynesian cousins in most parts of the country. The two Eurasian rats are blamed for destroying vegetation and eating the eggs and chicks of ground-nesting birds, and evidence suggests they wiped out the remnants of several species that made their last stands on smaller islands, including the South Island snipe (*Coenocorypha iredalei*) and the bushwren (*Xenicus longipes*).

"Rats are just the worst by numbers," van den Broeke says. "Then stoats are the second one on our hit list, mainly because they've got big territories and they're ferocious killers. Then all the other introduced, lovely, furry mammals: ferrets, weasels, hedgehogs, feral cats, feral dogs, and possums—the whole suite of them."

It's still not an easy thing for van den Broeke, who feels empathy for all furry animals, even when they're pests, not pets. But she has learned to extend that empathy to birds, for whom life has not exactly gotten easier since humans first discovered this avian paradise. In 2016, she was among the volunteers who helped to establish a new reserve outside Auckland, a 90-acre parcel that would be known as Matuku Link: "Matuku," for the threatened matuku-hūrepo that lives there (it's also known as the Australasian bittern, *Botaurus poiciloptilus*), and "Link" because it connected together several other conservation areas. The reserve would also pave the way for local reintroduction of the North Island kōkako (*Callaeas wilsoni*), a gray, pigeon-sized passerine that features prominently on New Zealand's $50 bill.

"That tells you something about their status," van den Broeke says. (Every denomination of New Zealand currency, from the dollar coin up, features one or more celebrated native species.)

As a volunteer with Matuku Link, van den Broeke assisted the team of scientists who relocated the kōkako, an experience that was personally transformative. After luring the birds into mist nets with a recording of their species' dawn call (which has been described as "haunting") the scientists and volunteers banded them for identification and moved them to their new home. "After that, you want to do everything possible to save them because you've looked them in the eye," van den Broeke says. "You've built this relationship with the bird and with the species, so you feel very protective of it and responsible for it. And then if that means you have to do more, you do more."

Today, van den Broeke is Matuku Link's full-time project manager, overseeing among other programs a team of volunteers who patrol traplines and maintain three hundred bait stations that target rats, stoats, and possums. She has reconciled herself to trapping the mammals that might threaten the kōkakos and other birds counting on her protection, and she's thinking hard about moving on to firearms.

"Almost fifteen years in, I'm still on the verge of deciding if I can pull the trigger," she says. "I'm not there yet. Once it's in a trap, or once it's dead, I can deal with it, no problem. And I know it's exactly the same thing, whether you put poison out, or a trap, or you shoot it. But pulling the trigger myself? I'm not there yet."

For van den Broeke's friend and neighbor James Russell, on the other hand, the calculations have always been simple: "You either have birds and no rats, or rats and no birds."

A professor of conservation at the University of Auckland, Russell is among the most respected figures in his field, a veteran of invasive predator removal projects not only in Aotearoa but on islands around the world. Along with British ecologist Tim Blackburn, he is also the author of the 2017 journal paper "The Rise of Invasive Species Denialism," which accused mainstream media outlets and a handful of academics of sowing unjustified doubt about the issue.

Russell does not, he hastens to add when we talk, entirely disagree with Mark Davis and other critics of invasion biology, who believe the field places too much emphasis on apparent nativeness. Indeed, some species introduced to New Zealand should be allowed to stay, Russell says.

"Nobody here is trying to go after roses or sheep because they're not native," he clarifies. Such species may have altered New Zealand's ecosystems, but they are also valued by people, either for economic or aesthetic reasons. But species like possums and rats, he argues, contribute nothing that might balance out the harm they do to native wildlife, vegetation, and agriculture. "Their impact is overwhelmingly negative. Their benefits are almost none. And we have the legitimacy, I would say, to eradicate them entirely from the country."

Thin, intense, and a bit disheveled, Russell speaks in rapid paragraphs and—perhaps because I have been thinking so much about birds lately—comes across as slightly avian himself. A lifelong bird lover, he grew up in suburban Auckland, in a family that was intimately acquainted with native species—specifically, the kākāriki, or red-fronted parakeet (*Cyanoramphus novaezelandiae*).

"It's a green parakeet with a little red hat," Russell says, explaining that his mother was a licensed breeder of the blue jay–sized birds. "So I just grew up with these cute little native parrots. And then I came to realize that they were extinct from most of New Zealand because rats have eaten them. And if you wanted to see them, you needed to go out to one of those islands that had either never had rats or where someone had, thankfully, removed the rats earlier and birds had been able to recolonize."

Thanks in part to his mother's parakeets, Russell had a love of birds that extended to the natural world in general. Studying biology in college seemed a natural fit, but when it came time to chart a career he set any romantic notions aside.

"I wasn't just wanting to do science for curiosity's sake," he says. "There's an infinite amount of science we could do for curiosity's sake, but I wanted to do solution-oriented science. So the culmination of that was, if you care about species extinctions, most of them have happened on islands. If you care about what's making species extinct on islands, it's invasive species. And the solution is, we can eradicate invasive species. So all roads led to here."

In graduate school at the University of Auckland, Russell studied the ecology and genetics of invasive rats, earning headlines for a brief 2005 *Nature* paper that documented the unlikely odyssey of a single male Norway rat, which he had fitted with a radio collar and released as part of an experiment on a small, rat-free island east of Auckland. The creature broke records by swimming a quarter mile to another offshore island—perhaps in search of a female—and then evaded Russell's increasingly frantic efforts to capture him for another eighteen weeks.

The case of the rat, nicknamed "Razza," had implications for projects to dispatch sparse populations of invasive animals, such as the first group of rats to reach an island, or the last furry survivors of a mostly successful eradiation program. It also inspired a children's book by University of Auckland English Professor and Māori memoirist Witi Ihimaera called *The Amazing Adventures of Razza the Rat*. The cover features a whimsical illustration of the heroic rodent hitching a ride with a friendly albatross, a bird that has itself fallen prey to rat invasion. (The real-life Razza was, fortunately or unfortunately, eventually trapped and killed.)

Today, with years of rat eradication success stories under his belt, Russell is a world authority on island conservation and an advisor to and cheerleader for the nationwide predator-free effort. He credits the movement's first successes, however, to people working outside the academy.

"Up until the 1960s, New Zealand was still a very colonial place," he explains, "and the academics and scientists were British imports, for the most part, and with them they imported what I would call 'continental' ecological mindsets."

On continents, he explains, there's some validity to the idea that predators are natural parts of ecosystems, and that excess predation will eventually come back into balance, rather than driving prey species to extinction. (The noteworthy exception is humans, who have been hunting continental species to extinction for millennia.) On islands, however, the balance of nature is a lot more precarious.

"In the 1960s," Russell continues, "these New Zealand wildlife rangers started to see that when you went to an island, it either still had its native birds or it had rats. The scientists from Europe were saying, 'No, no, they can coexist, they'll settle into a new equilibrium.' But these wildlife rangers were saying, 'There doesn't seem to be an equilibrium.'"

The watershed event was Big South Cape.

Black rats first appeared on that island—today known officially by its Māori name, Taukihepa—in the mid-1950s, and within a few years they were everywhere, to the chagrin of a handful of Māori families who had been visiting the otherwise uninhabited island for generations to harvest seabird chicks. These families, along with environmental groups and local representatives of the since-dissolved New Zealand Wildlife Service, worked together to poison the rats and tried in vain to drum up concern among the boffins in Auckland and the bureaucrats in Wellington, the national capital. But these experts doubted the rats were to blame for native birds' perilous state, arguing for habitat loss as the main driver of extinction. And anyway, long human experience had taught them that completely exterminating rats was a pipe dream.

According to a 2016 retrospective in the *New Zealand Journal of Ecology*, by 1964 at least two of Big South Cape's bird species had been extirpated, with others restricted to a redoubt at the island's southern end: "The impacts of rats . . . were already immense; there was absolute silence over most of the island." A team of ornithologists fanned out to capture surviving birds, which they evacuated to neighboring, rat-free islands or protected in secure aviaries (most, however, did not survive this well-meaning assistance).

Meanwhile, conservation groups began reporting surprising success elsewhere. In the early 1960s, poison eliminated rats on tiny Maria Island, a former World War II bombing range and a nesting ground for white-faced storm petrels, near Auckland. In 1988, they successfully used a new, slow-acting anticoagulant poison called brodifacoum to wipe out rats in just three weeks on 420-acre Breaksea Island, a forested rock that is part of the South Island's Fiordland National Park. Breaksea was soon recolonized by threatened native species, including the rare South Island saddleback (*Philesturnus carunculatus*) and knobbled weevils (genus *Hadramphus*), lumpy brown bugs that are among New Zealand's many legally protected invertebrates.

These successes, juxtaposed with the tragedy on Big South Cape, galvanized a change in attitude among scientists and officials, and eventually among regular New Zealanders. The last rats on Big South Cape were killed by aerial poison drops in 2006, but it is still considered the site of New Zealand's worst ecological disaster, on par with the infamous Exxon Valdez oil spill in Alaska.

"In a nutshell, that birthed the rat eradication industry, in which New Zealand leads the world," Russell says.

To date, rat eradication has worked on more than one thousand islands worldwide, including two hundred in Aotearoa alone. The global success rate now stands at more than 88 percent.

"It's all just variations of a theme," Russell says. "Islands popped out of the ocean and didn't have rats on them, rats and cats got there, and the birds went extinct. We have to do the same thing in Hawaii that we have to do in the Galapagos and in the Southern Ocean and in New Zealand. I've just been on a phone call with someone from the Caribbean, and it's all just the same. There's no magical different story. They're just islands. I can go to an island, and I can say, 'Hey, give me a million dollars. We'll get these rats off.' And that problem is solved."

Social Capital

Big South Cape covers about 3.5 square miles and has no year-round human population. Frigid South Georgia Island, a British territory that is the largest so far to be successfully cleared of rats, covers 1,362 square miles and hosts a winter population of fewer than a dozen hardy scientists. A quarter millennium after rodents first arrived there, it took four years, three helicopters, more than three hundred tons of poison and US$13.5 million to get rid of them, along with 4,600 peanut butter–coated bait sticks and a handful of sniffer dogs to make sure they were gone before conservationists declared victory in 2018.

Near the opposite end of the spectrum, New Zealand's main landforms, South Island and North Island, are the twelfth and fourteenth largest islands in the world. Together, they extend over more than one hundred thousand square miles and are home to more than five million people, with an additional three million visiting each year. Wiping out predators here is a far more ambitious goal than anyone has ever attempted, and if it's possible at all, it will mean mobilizing every man, woman, and child.

So far, though, that part seems to be going well.

It started, according to some sources, with a New Zealander named Les Kelly, who returned in 2007 after working in Australia's mining industry for twenty-five years. Soon after he got home, Kelly went "tramping"—an unofficial national pastime—and noticed far fewer native birds around than he remembered seeing and hearing as a young man. If predators were to blame, he wondered, why not just get rid of them? Kelly pushed the idea in his own network of conservation advocates, and—the details here are fuzzy—it eventually reached Sir Paul Callaghan, a physicist, popular science communicator, and frequent guest on Radio New Zealand, the country's equivalent of the BBC. At his last address before he died in 2012 at the age of sixty-four, after a long, public battle with colon cancer, Callaghan urged New Zealanders to take on the challenge of eradicating pest animals to save their native species.

"Let's get rid of the lot," he said, in a much-quoted lecture. "Let's get rid of all the damn mustelids, all the rats, all the possums, from the mainland islands of New Zealand. . . . It's crazy and ambitious but I think it might be worth a shot."

Callaghan was a beloved figure, and he found an eager audience for the idea that would come to be called New Zealand's moonshot. In 2013, a group of environmentalists and businesspeople founded the nonprofit Predator Free New Zealand Trust. Three years later, Prime Minister John Key announced that the government would fully commit to the cause, with a goal of eradicating rats, stoats, ferrets, weasels, and possums by 2050. (Cats, whether owned or feral, also pose a threat to native birds, bats, insects, and reptiles, but as more than 40 percent of New Zealand households include pet cats, explicitly targeting this particular invasive species was deemed a little too sensitive.)

"I make no apology for using fighting words," Lou Sanson, director-general of the Department of Conservation, said the following year. "We need to strike now if we are to win the war against invaders and restore our precious native species to health."

New Zealand government officials' evocation of "a predator-free Utopia" was a rallying cry at least as inspirational as US President John F. Kennedy's 1961 inaugural address ("Ask not what your country can do for you . . ."). Soon, New Zealanders of all stripes began asking what they could do for their beloved birds. The answer came easily: trap and poison possums, rats, and stoats.

"When the government put that flag in the ground, it created what I consider a social movement," Russell says. "In the last ten years, it's been massively transformative."

Today, the Department of Conservation, dozens of environmental organizations, regional governments, Māori groups, and universities are approaching the goal from different angles. Predator Free New Zealand Trust supports more than two thousand community groups with education, encouragement, and resources to kill predators in their backyards and neighborhoods. Its online shop sells traps to fit every need, from basic snap traps to the Bluetooth-connected Smart Trap by the New Zealand–based company Goodnature, which retails for NZ$199. There are also bird feeders and native bird and reptile posters, to help volunteers keep the ultimate goal in mind.

While official New Zealand is on board, it is broad-based buy-in from regular Kiwis that may make Predator Free possible. In the decade since the project began in earnest, hundreds of communities have rallied around it—in suburban Wellington, for example, where Predator Free Churton Park aims to "encourage native birds and lizards back into our gardens." Or, in the tiny South Island town of Brighton, where at a recent potluck dinner volunteers learned that the Brighton Trapping Project had recorded 929 kills over the previous year, including nearly six hundred possums.

"People talked about the flourishing of native plants in their native bush blocks, of seeing more native birds around, and even of the geckos on their land!" the group reported on Facebook. "It finally feels like we're making a difference."

Kiwis participating in Predator Free are seeing a difference not only in the relative abundance of various animals, but also in their own social capital. That's according to a 2022 study led by University of Auckland PhD student Rosie Gerolemou, who surveyed more than 1,200 Auckland residents about their involvement in community groups. She found that people who participated in local pest control efforts had more connections with friends and neighbors, along with greater feelings of trust and safety, than even members of other associations, such as those based on a religion, hobby, or sport. Even people who only trapped invasive animals on their own reported feeling more connected to others.

In 2021 and 2022, Alexandra "Ally" Palmer, a social scientist at the University of Auckland who specializes in human-animal relationships, interviewed several dozen New Zealanders involved with the effort in one way or another, including conservation workers and volunteers, Māori community members, researchers, government staff, and educators. While participants in the movement generally hoped that the 2050 goal could be met, they saw benefits even if the effort ultimately falls short.

Trapping is "a gateway drug to environmental understanding," one Māori conservationist told Palmer. "And it's such a simple thing. . . . What do I do? Do I stop driving my car? But a trap is simple. It gives you agency."

Others Palmer interviewed, like the people in Gerolemou's survey, placed value on the connections they made with other people.

"Getting to zero is cool," a conservation manager remarked, "but this project is about delivering social outcomes via conservation."

Dan Henry, a television producer by day, started Predator Free Miramar to tackle rats in his community of about ten thousand.

Known as the home of Sir Peter Jackson's film empire (the Wellington native directed *The Lord of the Rings* and *The Hobbit* trilogies), Miramar is a de facto island of its own, joined to the rest of North Island only by a narrow isthmus on which sits Wellington International Airport. Relative isolation made it a plausible target for invasive species elimination in the early 2000s, and between 2003 and 2006 the local government successfully rid the peninsula of possums. Rats, who so adeptly share humans' living quarters, were a tougher proposition, but Henry and a friend thought it was doable.

"We had a plan, but it was very basic," he says. "It was kind of like, 'Let's just give away as many traps as we can for free, and try and switch people onto this movement and get everybody to do their little bit.'"

To date, he says, the group has given out some 1,200 traps, many from a table set up at a local garden center, or to neighbors and friends after striking up a conversation at a dinner party. Word of mouth was key, he says, when it came to moving people from "What? That sounds gross" to enthusiastic, even competitive, backyard trapping. And while "birds in the bush" was the primary motivation for many trappers, others had more quotidian reasons for participating.

"We had people approach us because they'd had rats that had chewed through the dishwasher outlet hose and flooded the kitchen and caused thousands of dollars' worth of an insurance job," Henry says. Others had found rats living in their car's engine or heard them skittering in the walls at night—the sort of rodent problems with which any human in the history of civilization might sympathize. We, after all, have been living with rats far longer than have Aotearoa's birds.

Miramar residents' motives aside, the trapping campaign was remarkably effective in restoring native species, Henry reports.

Within a few years, fantails (genus *Rhipidura*) had returned to Miramar, and kākā (*Nestor meridionalis*), New Zealand's goofy, beloved native parrots, were spotted there for the first time in most people's memory.

"We had a Facebook group that took on a life of its own and became a really successful example of what a community can do together," Henry says. "We would post a lot of images of dead rats—whenever I caught a rat in my trap in my backyard, I would photograph it. As the rats became fewer, people were posting fewer and fewer of those, but they were posting other things—birds in their backyard, and little geckos that came in. They'd bring the washing in off the line and be folding the towels, and there was this lizard! If you go and look at our Facebook group and scroll through, chronologically, all the photos, you can see that shift from dead shit to cute animals."

Perimeter Control

The New Zealanders who are trapping rats and celebrating native gecko sightings have much in common with volunteers working to eradicate invasive species around the world, people like Jim Hurley and Susan Roth in Virginia or Premek Hamr in Ontario. But the broad base enjoyed by Predator Free is uniquely Kiwi—indeed, this may be the only country where such a movement could boast this level of engagement. New Zealanders, after all, freely admit to being "obsessed" with their native wildlife.

Since 1974, Radio New Zealand has begun every morning newscast with a native bird call, an institution that listeners once clamored to defend when the false rumor spread that it was ending. Such passion for Aotearoa's birds, as some have discovered to their peril, is not to be trifled with. In 2023, when a North Island brown kiwi (*Apteryx mantelli*) appeared to be receiving rough treatment from visitors at an American zoo, thousands of New Zealanders expressed their outrage through social media and an online petition, leading the zoo to shut down its $25-a-ticket "Kiwi Encounter" program and apologize for what had escalated into an international incident.

The same year, another hapless American ran afoul of New Zealand's conservation culture when he killed and ate a weka (*Gallirallus australis*), a protected bird the size of a chicken. The man and his teammate were competing on a *Survivor*-style reality TV show for USA Network, filmed in the South Island bush. The New Zealand Department of Conservation issued the producers a stern warning as the Americans did their best to smooth feathers.

"I made a mistake. It was shortsighted. It was foolish," the disgraced contestant said on camera. "What I did disrespected New Zealand, and I'm sorry."

Perhaps unsurprisingly for a nation of environmentalists, New Zealanders also trust science and scientists to a greater degree than citizens of almost any other country. According to a 2018 report conducted by Gallup, a third of survey respondents in New Zealand and Australia had high trust in science, compared to 18 percent worldwide. Ninety-three percent of New Zealanders had at least "medium" trust in science. That, perhaps along with the social cohesion that comes with being an island nation almost one thousand miles from the nearest continent, helped Predator Free win hearts and minds.

Trust in science and social cohesion were also crucial for New Zealand's COVID-19 pandemic strategy in the early 2020s, when the country staved off widespread infection for two years, longer than nearly anywhere else in the world. But northern Europeans are also pro-science and tightly knit, and their pandemic responses were more varied. It seems likely, then, that New Zealand's history with invasive species and its longstanding commitment to biosecurity primed its population to accept border closures, quarantine, and lockdown protocols.

The first in the world to enact a national biosecurity law, in 1993, New Zealand has strict rules about what travelers, immigrants, and importers can bring into the country, so as to keep them from introducing pests and pathogens that could threaten biodiversity and agriculture. Biosecurity preparedness is central to New Zealand's emergency management systems, and should the nation's borders be breached the Minister for Biosecurity and Food Safety can declare a "biosecurity emergency" that grants the government special powers, including to call on the armed forces for help.

Extending this framework to human health was not a major leap.

By the summer of 2020, New Zealand's status as a pandemic outlier had made then-Prime Minister Jacinda Ardern a household name around the world. For many in the West—particularly left-leaning Americans living reluctantly through the first administration of President Donald Trump—the distant land of Aotearoa appeared to be a progressive utopia. But social cohesion can also have a dark side, especially in a country that is so focused on its borders.

Over the past several decades, official New Zealand has worked to recognize and empower its Māori minority, which makes up nearly 20 percent of the population, and Māori values of stewardship are, at least in theory, incorporated into environmental policies like Predator Free. Meanwhile, the right of the white majority to continue occupying Aotearoa is not in dispute. But as the proportion of New Zealanders born overseas grew to one in four by the late 2010s, tension over more recent immigration has risen dangerously.

Before she became known worldwide for her handling of the pandemic, Ardern ran for office in 2017 on a Labour Party platform that aimed to cut immigration in half, to address the perception that newcomers, particularly Asians arriving in the Auckland area, were putting pressure on infrastructure and inflating housing prices. The question of who belongs in New Zealand was not new even then—in 2005, for example, the controversial Citizenship Amendment Act had rescinded the longstanding rule of automatic citizenship for anyone born in New Zealand, instead requiring that at least one parent be a citizen or permanent resident. But the issue has become all the more pressing since pandemic restrictions ended. Despite Ardern's campaign promise, in 2023, net migration into Aotearoa hit an all-time high of 110,000 people in a twelve-month period. By the following year, a newly elected conservative government had deemed such rates "unsustainable" and was working to tighten immigration restrictions.

The parallels between removing foreign animals to defend valued native species and barring foreign *people* to protect the interests of the native-born are both obvious and fraught—it's not a topic that anyone wants to talk about. But some in New Zealand are quietly concerned that anti-predator rhetoric may be tapping into, or even encouraging, nativist sentiments—especially when young people get drafted into the fight.

Like other good causes, from civil rights to climate activism, the movement to purge Aotearoa of invasive mammals has proven popular with educators, who see it as a way to get the next generation involved in environmental stewardship. When it comes to killing animals for the sake of conservation, however, a well-intentioned impulse to make things fun for children can occasionally lead in disturbing directions.

In 2012, the Uruti School, a small primary campus with about a dozen students in a remote part of New Zealand's North Island, held a fundraising gala that included a "best-dressed" competition. The evening, which raised NZ$8,000 for the school, was all fairly routine—except that the style icons on display were dead possums, outfitted by the schoolchildren in doll clothes and wigs. One fuzzy corpse wore a bridal gown, two were dressed as babies, and another appeared in pink princess garb, crowned with a sparkly tiara. One possum, its fur plucked to the neck for a bare-chested effect, was posed—with evident help from rigor mortis—as a boxer in a championship belt and gloves. Another lay spread-eagled on a towel as if sunbathing, clothed in a revealing bikini with a tube of sunscreen by its side.

"There was an amazing crowd and it was lots of fun," Principal Pauline Sutton told the press, which had a minor field day with the ghoulish pictures (the photos made the United Kingdom's *Daily Mail*).

The New Zealand Society for the Prevention of Cruelty to Animals was less amused—and positively livid five years later, when young participants in a possum-hunting contest staged by another rural school drowned baby joeys in a bucket (the youngsters had been found in their slain mothers' pouches).

Such distinctively macabre incidents are rare, but rural hunting competitions are not, and children often participate alongside adults in contests to shoot the most possums or trap the most rats. One South Island town stages an annual Great Easter Bunny Hunt; the 2025 event broke records with more than twenty-four thousand rabbits, hares, and other designated pests shot in twenty-four hours by thirty-nine teams.

"Let's get this out of the way, just in case PETA is watching: *Oryctolagus cuniculus* . . . is an introduced species in New

Zealand which threatens both native biodiversity and carefully-manicured lawns," one New Zealand journalist advised, before launching into enthusiastic coverage of the event. "Also, this story contains large amounts of ammo. I suggest you take a smoko, PETA, and sit this one out."

But animal rights advocates are not alone in questioning children's involvement in such activities. While most New Zealanders remain sanguine about removing non-native animals from the landscape, some have expressed concern about the effects of such "fun" on young people who may be just forming their moral sensibilities.

For a paper published in the *Australian Journal of Environmental Education* in 2019, a University of Auckland education researcher named Rajesh Ram surveyed and interviewed 171 Auckland teenagers about their opinions on biosecurity, the environment, and society. The youth were for the most part aware of, and concerned about, the non-native species that New Zealand has targeted for elimination, but their understanding proved to be simplistic and exaggerated. Asked what they knew about possums the teens' responses included not only "damages environment" and "ruins habitat," but "breeds quickly," "eats everything," and "sharp claws." (In truth, possums use their intimidating claws to climb trees, they generally stick to a diet of leaves, and they raise only one or two offspring each year.) Some of the young people proved somewhat conspiracy-minded about how invasive species reached New Zealand, surmising that "immigrants wanted to sell them," or they were "secretly shipped by foreigners" or "people from other countries like Jamaica."

"In this way the politics of national identity appears to have crossed paths with the issue of managing invasive species," Ram concluded.

Still, this study's young subjects should be forgiven for a lack of nuance, because Kiwis have historically been steeped in anti-possum rhetoric. According to a 2009 analysis by Annie Potts, co-director of the New Zealand Centre for Human-Animal Studies at the University of Canterbury, in Christchurch "the demonization of possums in New Zealand is overdetermined, extreme, and unhelpfully entangled in notions of patriotism and nationalism." Indeed, she found, it is "unpatriotic to question, let alone resist," the idea of possums as a public enemy.

In addition to potentially fueling xenophobia, the demonization of possums and other non-native mammals may also risk desensitizing young Kiwis to violence against animals in general, some observers worry. In 2021, nearly thirty-two thousand people signed a petition asking Prime Minister Ardern to end youth-focused trapping and hunting programs. They were joined by renowned British primatologist Jane Goodall, who argued in an open letter to New Zealand authorities, "Maintaining the highest standards of understanding, empathy and compassion shapes [young people's] view of all living things. This will benefit who they are today and the communities we have tomorrow."

Indeed, a long-standing body of research links empathy toward animals with prosocial behavior, while youthful animal cruelty correlates with an adulthood propensity for domestic violence, criminality, and aggression toward people. In his widely cited 2005 book *Children & Animals: Exploring the Roots of Kindness and Cruelty*, the American psychologist Frank R. Ascione cites studies that, together, "suggest that animal abuse may be characteristic in the developmental histories of between 1 in 4 to nearly 2 in 3 violent offenders. . . . Clearly a significantly higher rate than what we would find in the general population of non-offending adults." To be sure, animal abuse has not been determined to *cause* later criminality, but it's clear to psychologists and laypeople alike that encouraging children to be cruel to animals is unlikely to encourage empathy.

Increasingly, however, even supporters of the Predator Free initiative have become aware of these and other ethical traps.

Ally Palmer, the University of Auckland social scientist, is a somewhat ambivalent trapper: a vegetarian who loves animals, she sees the practice as sad, yet necessary for the sake of New Zealand's vulnerable ecosystem. Recently, she explored the phenomenon of New Zealand conservationists sharing photos and graphics depicting dead "pest" animals online, for a paper she titled "Digital Animal Deathscapes." Such posts, she argued, served to "perpetuate the public un-grievability of 'pest' animals," despite the feelings of sadness that many who do the killing might express privately.

Images of dead animals, which can sometimes be graphic or flippant—think of those dressed-up possums—are common on the Facebook pages of some local conservation groups or in private chats on the messaging platform WhatsApp. Palmer has observed that these pictures are used not only to seek support and urge others to get involved in predator control, as in Dan Henry's Predator Free Miramar Facebook group, but also to overcome feelings of moral disgust and pity toward target animals. This, in turn, may help promote what Palmer terms "social license," or the passive acceptance of killing pests.

"People need to see that we're getting more birds because of animals dying, which is brutal, but it's maybe a better understanding of that situation," one conservation manager told Palmer.

But perhaps it's possible—and even helpful—to acknowledge the sense of grief that animal lovers may feel when they kill for what they believe is a greater good, Palmer argues. Killing *should* present us with a moral dilemma, and if it's actually easy for us, if it's something we're comfortable with—well, we may have lost the plot.

During the 2010s, amidst the first flurry of patriotic excitement after Predator Free 2050's official launch, media reports were rife with declarations that possums represented "an ecological nightmare," rabbits were "evil," and rats considered New Zealand birds' eggs a special delicacy. Predator Free New Zealand Trust, the effort's highest-profile nonprofit, announced itself with a cheeky logo of a kiwi in silhouette, clutching a dead rat in its talons (the rat had Xs for eyes, lest anyone misunderstand).

But in 2023, the organization commissioned a new logo from Māori graphic designer Tyrone Ohia. In this one, the kiwi stood alone.

"We're moving on from focusing on death and killing in the name of conservation," the group explained, announcing the new logo on its website. "[W]hile it's a valid reality we have to face up to, we want to focus on the outcomes of predator control: saving the birds, bugs, bats and lizards that deserve to thrive in Aotearoa."

The new logo was part of a larger reframing, says Jessi Morgan, the group's chief executive, in a call from the group's Wellington headquarters.

"We just thought we were ready," she explains. In addition to shifting the focus from a dead rat to a live kiwi, the organization has also moved away from militant language, the idea that "we'll build an *army* of volunteers, and we need to *fight* this together. Actually, what we're trying to do here is create a land where our native species can thrive and flourish, and so can our people."

If the mobilization metaphor is looking less helpful, the idea of introduced predators as "the enemy" will also have to go, Morgan adds. "One of the things that we try to do really clearly in our messaging is not villainizing the animals. The animals are sentient. They deserve to be treated with respect and not suffer.

And also, it's not their fault that they're here, they're beautiful animals, but they're just in the wrong place. Humans introduced them, and so we need to make sure we deal with the removal of them in a humane way."

Morgan spent more than a decade in the business world before getting involved in environmental nonprofits, and she brought along a sense for the importance of careful messaging. If it's to succeed, New Zealand's predator control movement must be nonpartisan, the sort of big tent that can accommodate Kiwis of all dispositions and walks of life. In that way, high-profile pest hunts can be a problem—but the culture from which they emerge also cannot be dismissed entirely.

"Possums have been a pest for a long time, especially in an agricultural setting," notes Morgan. "We have to understand that people have grown up in those rural environments shooting possums out of trees, and it's very normal for New Zealand farmers to have to control possums on a farm. We need to understand those value sets but also try to educate or change the way we think of these animals—we do need to treat them with respect, or we'll lose our social license to operate. If people think they're suffering, it's unacceptable."

An Ocean Away

In 2019, Emily Major arrived in Christchurch, New Zealand, and found herself in what might as well have been an entirely different world.

"Coming from Canada, I just thought of New Zealand as Middle Earth, a place that had beautiful mountains and everything," says Major, who grew up in a working-class family in Ontario, where—like Joseph Banks—the only "possum" she had ever encountered was the Virginia opossum.

Major knew, vaguely, that New Zealand had unique animals, and she was excited to learn more about them when she enrolled at the University of Canterbury as a candidate for one of the world's first doctorates in human-animal studies. Her research would take place at the thorny intersection between ecology and culture—one that, in New Zealand, proved thornier than she could have guessed.

"I was really interested in the concept of 'speciesism,' and how different pest species can be treated differently based on where they are," explains Major. Her thesis advisor was Annie Potts, who had researched anti-possum rhetoric in the years before Predator Free became a national movement. Hoping to help Major narrow her topic, Potts explained the situation with possums. Major, a vegan who keeps pet rats, was aghast.

"I just thought, 'What? You've got to be joking me,' and so I started doing more reading," Major recalls. "New Zealand is really progressive when it comes to humans, but it's almost the complete opposite when it comes to animals. Everything is so regimented and controlled. It's incredibly conservative." Major, who considers herself an "academic activist," was distressed to learn that many New Zealanders see pest control not only as an environmental necessity but also a patriotic duty. Indeed, trapping seemed all but a requirement for citizenship—especially among white people whose colonial ancestors had imported most of the invasive species to Aotearoa in the first place.

"It's kind of wanting to cement your place within the nation," Major says.

Focusing her ethnographic research on possums, Major interviewed New Zealanders with minority views, including animal welfare advocates and people who had adopted orphaned joeys (the latter often struggled to find veterinarians who would provide medical care). Like the very idea that a possum could be a pet, Major's line of inquiry proved difficult for many Kiwis to accept.

"I got a death threat once, which was just ridiculous," Major says. "It was scary. This guy found out what my research was about and messaged me and tried calling me several times"

People who attended her lectures spoke over her, trying to correct what they saw as her mistaken ideas. On Reddit, someone at the r/NewZealand discussion board posted an interview she had given. Commenters ripped it to shreds, questioning the legitimacy of her doctoral program, her funding, and even her sanity.

"Possums eat baby birds," one declared flatly. "You do you but I'm looking forward to all the possums being dead."

"Some people seem to be so disconnected from reality," sighed another. "What are we supposed to do, rehabilitate possums and teach them to eat tofu?"

With her PhD in hand, Major returned to Canada in 2023, leaving behind a dissertation that directly challenged New Zealand's national self-conception. Its title, *Possums are as kiwi as fish and chips*, refers to a local culinary institution that, like the majority of white New Zealanders, originated in Great Britain. In the words of one of her subjects, an animal rescuer named Laurie: people, possums, and British food "were all transplanted here, and we are all a part of the New Zealand experience."

Whether or not possums are here to stay, they have, like the widely observed tradition of Friday night fish and chips, certainly brought the country together, albeit in a way that can seem alienating even to the citizens of other Anglosphere nations. There are, after all, few things more ingrained in New Zealand's culture than its all-in approach to wiping out invasive species.

Interlude

SQUASH IT!

For years, the spotted lanternfly has been on its way.

In 2014, the gaudy planthopper (*Lycorma delicatula*) was detected outside Philadelphia, presumably after arriving from its native range in Southeast Asia as a stowaway on a container ship. In the summer of 2020, New York, still in the grip of the COVID-19 pandemic, announced an infestation on Staten Island. Citizens across the Northeast were put on alert, urged to kill any lanternflies they saw, and then to report them to their local authorities.

The stakes were purportedly high. While the lanternfly's preferred food source is tree-of-heaven (Ailanthus altissima, itself a despised invasive), in that tree's absence the insect is far from choosy, sucking sap from just about anything, including valuable food crops like grapevines and apple trees. Penn State University researchers have estimated that spotted lanternflies could cost Pennsylvania's agricultural sector as much as $554 million a year and mean the loss of five thousand jobs. Faced with such staggering impacts, the state agriculture department's message became, "Kill it! Squash it, smash it . . . just get rid of it." And though the Keystone State was ground zero, it's now far from alone on the front lines. Neighboring New Jersey, for example, has placed all twenty-one of its counties under quarantine, while billboards along the Garden State Parkway urge people to "Join Jersey's Stomp Team."

Stomping campaigns have been greeted with both public enthusiasm—one app developer created Squishr ("Make squishing the SLF fun")—and ambivalence, as think pieces wonder whether stomping lanternflies is truly an effective biosecurity effort or just another way to vent our baser impulses. Either way, over the past decade, *L. delicatula* has become a marquee invasive species.

It always takes a while for the hottest touring acts to make their way to the flyover states, but in the summer of 2024, I hear that spotted lanternflies have arrived in Toledo. In September, a local TV personality shares a video of them swarming downtown "by the thousands." Word is people are fighting them with leaf blowers. And while the Ohio city is still an hour's drive from my home in Michigan, it's close enough for me to finally see this infamous insect in person.

Picking a patch of green on the map, I wind up at Glass City Metropark, which sits on reclaimed industrial land along the south side of the Maumee River. As I step out of the car, the wind off Lake Erie picks up, and since the park is nearly empty of pedestrians I pop into the Museum of the Great Lakes, where a clerk says he sees at least one lanternfly out front most days.

"You're supposed to report them, but I don't know where," he says, shrugging. "I do stomp them, though."

Outside, concrete paths wend between garden plantings and recreated native prairies, and the whole place buzzes with insects—crane flies, bees, startled grasshoppers, and shiny black crickets. In the vegetation they sing the last desperate strains of their summer song. No lanternflies here, however.

"Just sit on our patio for ten minutes, you'll see one," a man behind the counter at the park's concession tells me. "They'll fly right at you!"

Everyone I meet in Toledo is aware of the lanternfly plague, with most avowing that they squash them on sight—or try to. Some believe the bugs are getting smarter and harder to catch. Still, I have yet to find one. Perhaps I'm too late, and they've moved on as quickly as they arrived.

Then, finally, I hear about some loft apartments near the Toledo Mud Hens' baseball stadium where lanternflies were recently spotted swarming. I hop back in the car and head over, and it's with a little thrill that I spot the first one within seconds of parking. It's smaller than I thought it would be—I guess the pictures in the news neglected to include anything for scale, but this bug is no more than an inch long from the tip of its snout to the end of its abdomen. Its soft beige wings, decorated elegantly with black polka dots, argue for the appropriateness of the insect's Latin name, *delicatula*.

It's also dead.

Then I notice more, all dead. Dozens of them. A few are seemingly frozen in place, perhaps killed with an application of insecticide by the building's management. But most are simply squashed, their black legs grotesquely akimbo, the pretty scarlet of their lower wings exposed. The people of Toledo have done their duty, and this, apparently, is what remains of the swarm that struck their city.

8

CONSERVATION, COMPASSION, AND CUTENESS

Perhaps it's the fluffy, expressive tail, the curious mien, or the paws they use just like tiny hands, but many humans simply can't resist a squirrel. Some time ago, my family hosted a pair of visiting Japanese high school students. On a tour of Washington, DC, they found the Eastern greys (*Sciurus carolinensis*) frolicking on the grounds of the US Capitol adorable and hilarious—and much more appealing than any of our nation's marble monuments. But Hiroko and Miu were far from the first visitors to fall for grey squirrels. Indeed, for much of the 19th and 20th centuries people around the world sought to bring this charismatic species home to their own countries.

In 1948, for example, two pairs captured in Washington—perhaps distant relatives of the Capitol lawn denizens that charmed our guests—arrived in northwestern Italy. Sources differ as to whether these Eastern greys were gifts from the American ambassador or souvenirs brought back by a returning Italian diplomat, but either way they were warmly welcomed and found a comfortable new home at the *Palazzina di Caccia di Stupinigi*, a sprawling Baroque palace surrounded by a large park.

The rodents were not Stupinigi's first exotic denizens. For a time in the 19th century, the Savoyard kings who lived there kept a menagerie that included kangaroos, ostriches, and an Asian elephant named Fritz. Open to the public after World War I, Stupinigi soon became a premiere tourist attraction, with guidebooks highlighting the abundant wildlife—including foxes, weasels, and storks—in its protected park. The bushy-tailed American quartet fit right in.

Within a few decades of the first squirrels' arrival, however, the population of eastern greys had expanded beyond the park's boundary, and by 1997, they occupied some ninety thousand acres of Italy's Piedmont region. Officials began to fret that the industrious rodents would eat their way through the area's commercial poplar plantations and valuable hazelnut crop (the Piedmont, incidentally, is the birthplace of Nutella). Conservationists also feared that as the interlopers spread through wild places they would damage woodlands by stripping bark from trees (no one was able to explain why this alleged habit wasn't a problem in the forests of North America). But the most pressing—and most troubling—question was what would happen as *S. carolinensis* encountered its native cousin, the smaller, shyer *S. vulgaris*.

Europeans once considered red—or "common"—squirrels "a fast-breeding vermin species no different from rats, mice and rabbits," as the British environmental historian Peter Coates put it in *Squirrel Nation*, his 2023 book about sciurine controversy in the United Kingdom. But now reds were valiant underdogs, vulnerable not only to competition for habitat and resources but to squirrelpox, a virus harmless to greys but frequently fatal to reds. Indeed, while *S. vulgaris*'s range still spans Eurasia from Portugal to Japan, by the 1990s its population had collapsed in England, where hardy American greys made inroads as pets during the Victorian era. Were the Piedmont greys to spread further, conservationists feared, it might be all over for the tufty-eared rodents.

"A colonization of the entire Alps in the middle term and of a large part of Europe in the long term is predicted, potentially threatening the survival of the red squirrel in the continent," warned Piero Genovesi, a scientist at Italy's National Wildlife Institute, a quasi-governmental research body, in a paper presented at the University of California's annual Vertebrate Pest Conference, in 2000.

Given the urgency, National Wildlife Institute scientists felt justified in organizing a campaign to eradicate grey squirrels from Italy, where their population in the late 1990s was estimated at more than six thousand. The greys would be captured in traps baited with corn, then killed with an overdose of the anesthetic gas halothane, all under the supervision of a veterinarian. The team piloted their project at the Royal Park of Racconigi, another former House of Savoy property.

"The adopted procedure of euthanasia significantly reduced animals' stress; squirrels reached unconsciousness in less than a minute, and could be euthanized on the field, with very limited manipulation," reported Genovesi, the program's coordinator, after the successful dispatch of nearly two hundred squirrels from the stately home's grounds.

Though the park was closed during the campaign and the killings took place out of the public eye, word got out anyway, and many Italians were incensed. Alongside photos of cute squirrels, newspapers compared the eradication program to Nazi gas chambers. Giorgio Celli, host of the popular Italian television program *Nel regno degli animali* ("In the Animal Kingdom"), spoke out against the slaughter. Three animal rights groups sued, and Genovesi and a colleague found themselves in court, charged with illegal hunting and cruelty to animals. In December 1999, a judge ruled against the scientists, issuing a fine that amounted to about US$2,700 in today's dollars, and sentencing them to twenty days in jail (the jail sentence was later commuted and the fine halved). The scientists appealed, and in 2000 a higher court reversed the judgment. But by then it was too late. Containing grey squirrels now appeared to be impossible.

"[A]s a consequence of the suspension of the eradication campaign, the species has significantly expanded," wrote a regretful Genovese. "The opposition to the eradication of the grey squirrel in Italy represents an example of the growing conflicts between radical animal rights movements and conservation, and may result in dramatic consequences at a continental scale."

If invasion biology is a values-based science, as observers like Mark Davis have noted, the squirrel incident shows that its values are far from universal. Indeed, they may often be directly opposed to the values of the public at large, who tend to sympathize with individual creatures—particularly charismatic ones—rather than more abstract and hard-to-define concepts, like the future of a species or the integrity of an ecosystem.

While the threat of jail time is a rare complication, the conflict continues to plague—and baffle—scientists, conservationists, and land managers who are working to mitigate the effects of what one might call the cuter invasive species.

In a way, however, the Italian scientists were lucky. At least they weren't taking on cats. Even New Zealand has hesitated when it comes to cats—and, as some conservationists have discovered, with good reason.

Claws Out

Descended from Old World wildcats (*Felis silvestris*), domestic cats (*Felis catus*) began living alongside people as long ago as 10,000 BCE, welcomed by prehistoric farmers as allies against the mice and rats that ate and despoiled their grain. Later, cats provided invaluable service aboard rodent-plagued ships, which eventually carried them to every continent where humans live, as well as to thousands of far-flung islands.

Wherever they find themselves, these adaptable predators have quickly gotten down to the grisly business that millions of years of evolution taught them to do happily and efficiently. By one estimate, non-native cats are responsible for 14 percent of modern bird, mammal, and reptile extinctions worldwide, nearly all on remote islands with few native predators. Today, free-ranging cats take the blame for ecological havoc, even as they maintain a special relationship with a sizeable proportion of humanity.

Peter Marra, though slightly allergic to cats and perhaps more of a dog person anyway, started out as a neutral party.

"I definitely had no negative or positive feelings about cats," avows Marra, dean of the Earth Commons, Georgetown University's recently launched environmental institute. Indeed, for a time in college he amicably shared an apartment with a blameless indoor cat named Tuchus.

"But then I started to really look," he says. "As any researcher does, you publish one paper and then it starts to open up more and more doors, and you start to ask questions: Well, what do we really know here? How big of a problem is this?"

Marra's meandering route to those questions began in the 1960s, when he was a lonely latchkey child in suburban Connecticut, the youngest son of a single, working mother. Arriving home from school to an empty house, he would head out the back door and amble through the woods on his own.

"I used to be a free-range kid, just totally free range," he says, noting that such a childhood would not be possible in the 21st century. "At like five and six, I could do what I wanted to do, so I would wander down to the local nature center and be a pest."

Fortunately, the staff at the Westport Nature Center were tolerant of curious little boys, and they let Marra watch as they carried out research projects around the sixty-two-acre preserve, including capturing and banding songbirds to track their behavior, migration, and populations. Once, they let him hold a black-capped chickadee in his own small hands. A sense of wonder has stuck with him ever since: "It's one of the few things I remember clearly from my childhood."

A songbird is an exquisite, elusive thing, a creature we spy on from afar or catch in glimpses, and to capture and hold one can feel like crossing a sacred boundary. A chickadee in the hand is more delicate and vulnerable than even the most avid bird-watcher might imagine: about the length of a man's middle finger, it weighs a little more than a US quarter. When under stress, its tiny heart can beat nearly one thousand times a minute.

For Marra, this intimate moment, and other such encounters with the natural world, sparked interest in all kinds of wildlife, including turtles, salamanders, frogs, and mammals. As a teen, he had a pet raccoon, an orphan he rehabilitated and later released back into the woods. "But birds, I was always definitely drawn to them a bit more," he says.

In high school, Marra got involved with the Connecticut Audubon Society, banding birds on his own, and after earning a PhD in ecology at Dartmouth he spent twenty years as director of the Smithsonian Migratory Bird Center, housed at the National Zoo in Washington, DC.

It was there that he started to think differently about cats.

Soon after arriving at the Smithsonian, Marra created Neighborhood Nestwatch, a program to help researchers understand the impact human activity had on birds in a human-built environment. The DC-based citizen science project, which remains active although Marra has moved on, would also give regular people an opportunity to make intimate connections with the natural world. "In everything I do, it's really important that I try to engage the public to get more people interested in nature," Marra says.

Participants in Neighborhood Nestwatch were trained to catch birds, mark them with color-coded leg bands, and observe them in their own backyards. They learned how to find nests and monitor the success of fledglings, and they recorded which birds came back the following year, using the colored bands to identify them as individuals. ("It automatically creates a bond," Marra notes.)

Nestwatch also provided Smithsonian researchers and a rotating cast of graduate students with ready-made study sites to explore particular research topics. One year, they placed tiny transmitters on the backs of gray catbird nestlings, for a study of post-fledging survivorship that would be published in a 2011 issue of the *Journal of Ornithology*.

The result was an inventory of tiny tragedies.

Gray catbirds (*Dumetella carolinensis*) are a common backyard species, found throughout most of temperate North America in the summer and year-round on the East Coast, from Massachusetts to Georgia. Named not for any affinity with cats (far from it) but for their distinctive mewing call, catbirds prefer thickets and dense shrubs, the sort of edge habitat that's common in areas disturbed by human development. But convenient though it may be, living near people has a price.

The team monitored sixty-nine fledglings in three neighborhoods and recorded forty-two deaths, with more than three-quarters caused by predation. In some cases, it wasn't clear which predator had killed these young catbirds. The researchers witnessed one getting eaten by a black rat snake (*Pantherophis* spp.) and another snatched by a red-shouldered hawk (*Buteo lineatus*). Six were murdered in plain view by domestic cats. Other culprits could be assumed: "[F]ledglings found with body damage or missing heads were considered symptomatic of cat kills [and] those found cached underground of rat or chipmunk predation." (Yes, like most rodents, the unassuming Eastern chipmunk, *Tamias striatus*, is an opportunistic omnivore.)

While it's an unavoidable fact of nature that some number of baby birds are going to feed some number of predators, nearly half the known hunters in the Smithsonian study were cats, an animal that, in probably the majority of cases, does not *need* to hunt birds to survive. While instinct may drive them to kill, even stray cats are often fed by people. Moreover, free-roaming cats, like any species deemed invasive, represent a human-made problem, one that Marra came to believe we are also obligated to solve.

"Areas where there were cats were what we call 'demographic sinks,' where the adult birds can't produce enough young to maintain the population size," Marra says. "That got me starting to really think about cats and really trying to put together some information on the cat issue, because it always seemed to me that this was something that people could change. This is something we can do something about, unlike climate change. Little did I know that we would probably save the climate before we would get cats off the landscape!"

It was a small study, but the Smithsonian catbird paper got outsized attention.

"Cats Are a Bird's No. 1 Enemy," declared a *New York Times* headline, above an article that quoted Marra comparing cats to "gypsy moths and kudzu—they cause major ecological disruption."

That sort of rhetoric raised the hackles of a group called Alley Cat Allies, or ACA, a national nonprofit based in Bethesda, Maryland—right in Marra's own figurative backyard. Founded in 1990, ACA is one of the most prominent advocates of a strategy known as "trap-neuter-return" (or "TNR," in this initialism-loving community). The idea is to live-trap stray cats, bring them to a veterinarian for sterilization, mark them with an ear notch to indicate they've had the procedure, and—unless they seem friendly enough to become pets—release the desexed cats to go about their business. In theory, free-ranging colonies will naturally diminish over time, as older cats die without leaving any heirs to take their place on the streets. Research is mixed on whether this happens in practice, but regardless, resisting any lethal population control is central to ACA's mission.

Shortly after the Smithsonian study began circulating, a post on ACA's website broke down the paper's math, arguing that while its abstract (and journalists who reported the findings without reading past that section) highlighted cats' culpability for "47% of *known* predation events," the details revealed that, when *unknown* predators were counted, felines could truly be held responsible for just nine dead fledglings, out of an original sixty-nine chicks, or 13 percent. The paper was thus "bogus" and "irresponsible," the group declared.

The controversy caught Marra off guard. To him, the results of this early study had only made the matter more urgent. If cats were killing even 13 percent of catbird chicks in this one neighborhood, what was their broader impact on avian life?

In 2013, Smithsonian postdoc Scott Loss; Tom Will, a scientist at the US Fish and Wildlife Service; and Marra coauthored a paper in the journal *Nature Communications* that attempted to clarify, once and for all, just how bad cats are for birds. From a systemic review of previous publications on the topic of cat predation around the world, Loss et al. estimated that free-ranging cats kill as many as four billion birds annually in the United States, along with perhaps twenty-two billion mammals. The figures, they wrote, would make *F. catus* "the single greatest source of anthropogenic mortality for US birds and mammals."

Few humans seem to have lamented the mice, rats, and even chipmunks—those adorable little butchers!—that made up the enormous mammalian death toll. Birds, however, have a more powerful constituency.

In the United States and much of the developed world, those devotees include not only conservation nonprofits like the Audubon Society but millions of passionate bird-watchers, the sort of people who compile life lists, maintain backyard feeders, and subscribe to *Living Bird* magazine—and that's as many as a third of US adults, according to a 2024 survey by the Cornell Lab of Ornithology. The 2013 paper was "a groundbreaking study," declared the American Bird Conservancy, one that organization hoped would lead cat owners to begin treating their pets more like dogs—that is, kept indoors or on a leash.

Keeping pet cats indoors is no longer especially controversial (although Fluffy may have her own thoughts about the leash). But Marra and his colleagues soon found themselves facing accusations that their research would lead to the killing of feral cats. Many of these free-roaming felines are tenderly looked after by human caretakers, who feed and protect them despite perhaps never receiving a purr of thanks. But estimates of the feral cat population run as high as eighty million in the United States alone, and Marra's research suggested they were guilty of far more predation than pets were. Sensing a target on the animals it existed to protect, Alley Cat Allies issued another news release, declaring the latest paper "a veiled promotion by bird advocates to ramp up the mass killing of outdoor cats."

"This study is part of a continuing propaganda campaign to vilify cats," wrote Becky Robinson, the group's president and cofounder, noting darkly that another postdoc at the Migratory Bird Center had recently been convicted of attempted animal cruelty for trying to poison free-ranging cats at her DC apartment complex. (For what it's worth, Marra neither knew nor approved of these alleged extracurricular activities. His erstwhile colleague, meanwhile, has maintained her innocence.)

Writing on his blog *Vox Felina*, TNR advocate Peter Wolf dissected the paper's math, noting that the entire breeding population of North American land birds has only ever been estimated at a few billion. "Were the authors' estimates even remotely accurate, birds would have vanished from the U.S. long ago," Wolf wrote. "This was, in other words, classic junk science."

Marra reacts with exasperation to such criticisms, which continue to crop up in discussions of his bird mortality estimates.

"It's not inconsistent at all," he says. "And if you understand biology and science, it makes total sense. It's called population dynamics." In reality, he explains, wild bird populations fluctuate dramatically over the course of a year; their numbers may grow five- or tenfold as young are hatched and fledged, but because of a plethora of causes, ranging from disease to window strikes, a majority of this potential next generation won't see a second spring—and even fewer will if cats are eating them, too. "The three billion is really out of what *could* be."

As passionate as his critics, Marra was no less able to back down. Instead, he coauthored a book, *Cat Wars: The Devastating Consequences of a Cuddly Killer* with writer Chris Santella. Published in 2016, it laid out the case against free-range felines and called for getting cats off the streets—alive if possible, but ultimately, "by any means necessary":

> In high-priority areas there must be zero tolerance for free-ranging cats. If the animals are trapped, they must be removed from the area and not returned. If homes cannot be found for the animals and no sanctuaries or shelters are available, there is no choice but to euthanize them. If the animals cannot be trapped, other means must be taken to remove them from the landscape—be it the use of select poisons or the retention of professional hunters.

As the public face of a perceived war on cats—one that theoretically might one day involve armed counter-feline forces—Marra continued to weather fury online, and even in person. *Vox Felina* publicized his book tour, and opponents came to his talks, picketing in front of venues and distributing pamphlets that denounced his work. At Ohio State, one protestor was so disruptive that organizers called the police. A meeting with a local humane society devolved into accusations.

"I met with them to see if we could talk about common ground," Marra says, "and they just attacked me. They were not interested in having a conversation. I mean, the things they said! 'How could you be this way? How could you kill cats?' I've never killed a cat! I would find it a very hard thing to do. Again, I love animals!"

A Moral Panic?

Like the Italian squirrel debacle, the cats-vs.-birds controversy always seemed tailor-made for the popular press—and, in fact, the Smithsonian research and the *Cat Wars* book inspired multiple magazine features, which used the clash of personalities to punch up a compelling narrative. But in scientific journals, the drama persisted even after *New York* and *The Atlantic* had moved on.

In 2010, a bestseller called *Merchants of Doubt*, by historians of science Naomi Oreskes and Erik M. Conway, had revealed how a handful of scientists acted as useful skeptics for special interests working to sabotage progress in public health and environmental protection. Though it didn't deal specifically with non-native species, the book made a splash among invasion biologists. It was cited more than once, for example, in the flurry of responses and counter-responses to Mark Davis's infamous *Nature* paper.

If the scientists guilty of "invasive species denialism" were manufacturing doubt, so were *Vox Felina* and Alley Cat Allies, argued Loss and Marra, who reunited for a 2018 letter in *Conservation Biology* in which they compared "free-ranging cat advocates" to "cigarette and climate-change fact fighters." A few months later, the pair joined their 2013 collaborator Tom Will and Travis Longcore, an urban ecologist then at the University of Southern California, to respond blow-by-blow to a *Vox Felina* post by Peter Wolf. Published in the journal *Biological Invasions*, the article represented a surprising descent, given that blogs do not generally carry the gravity of peer-reviewed science. It spoke to the frustration the authors felt over being cast as one side in a debate, rather than as scientific experts whose judgment ought to be respected—and acted upon, before the crisis they foresaw became a certainty.

"Free-ranging cat advocates propagate misinformation about the ecological impacts of cats . . . to overturn policies that would allow removal of cats to achieve biodiversity management objectives," they wrote, noting that Loss et al. (2013) was "well-received in scientific circles, having been cited >320 times as of July 2018 according to Google Scholar, with no instances of negative criticism."

By this time, however, criticism from fellow scientists was starting to trickle in.

Responding in *Conservation Biology* the next year, a multidisciplinary group of academics in the United States and Australia disputed that cat advocates represent a "special interest" on par with Big Oil. In "A moral panic over cats," they pointed to scientific evidence complicating the idea that loose cats were an unequivocal ecological negative—for instance, studies of predator-prey dynamics on islands have shown that eliminating cats can actually reduce the breeding success of seabirds if it leads to an increase in the population of rodents.

But the authors disputed more than just the facts of the case.

"[T]he harming of sentient, sapient, and social individuals, such as cats . . . requires strict ethical and scientific scrutiny," they wrote, urging caution against "conservationists overgeneralizing their science and losing their moral compass."

William Lynn, the paper's first author, is a research scientist at Clark University in Worcester, Massachusetts, and the founder of PAN Works, a think tank and consultancy that advocates for considering animals "both as individuals and as active members of their broader ecological and social communities." Trained as a geographer, he has spent much of his career exploring the ethics surrounding humans' interactions with animals and the natural world.

Like Marra, he fell into the cat controversy partly by chance.

"I was asked to speak at the last moment at a conference on outdoor cats, and someone did a data dump of fifty scientific articles, and I started reading them," he recalls in an interview from his home in rural western Massachusetts. "I realized not only how bad the science was, but how absent the ethics was." Arguments for free-roaming cats were little better, says Lynn, who finds the evidence for "trap-neuter-return" likewise unconvincing. "Those kinds of flawed claims about causation and prediction just can't be trusted on either side." Conversations with likeminded colleagues led to the "moral panic" paper: "We wanted a more nuanced, ecologically and ethically grounded argument. We wanted to break the orthodoxy."

Orthodoxy, a word synonymous with "dogma," may seem anathema to the spirit of free inquiry that nonscientists tend to associate with science. But although rarely discussed, orthodoxy is foundational to many scientific fields, not least conservation, ecology, and invasion biology. And in Lynn's view it was more perilous for being unexamined.

"What we call 'scientism,' the ideology of science, hides those latent values," he explains, "but they get expressed through different visions of how nature ought to be."

There's nothing wrong with a scientist having values, Lynn hastens to add, noting his own strongly held belief in protecting and restoring indigenous biodiversity, which informs—among other projects—his active support of wolf reintroduction in North America. "But I also *know* that that's a value, and I'm willing to argue for that from a values perspective."

Acknowledging how different people weigh values differently, and how these values may inform the production of scientific knowledge, Lynn says, could allow scientists to debate controversial topics in better faith, especially when trade-offs are inevitable.

Take animals, for example. Many of us are inclined to see the chickadee that visits our birdfeeder or the cat that meows at our back door as individuals with a moral weight that, if perhaps not as great as a human's, at least makes them worthy of our consideration. Others may instead take the anthropocentric position that animals should be of interest to people mainly as resources (chickens, deer) or threats (wolves, mosquitos). This latter view long dominated in the West, where it was reinforced by the biblical teaching that God granted man dominion over the rest of His creation. (Although, as author Emma Marris revealed in her 2021 book *Wild Souls: Freedom and Flourishing in the Non-Human World*, there is also a history of Christian communities considering animals as moral agents in their own right—in medieval Europe, for example, authorities sometimes tried pigs for homicide or brought lawsuits against insect pests.)

Today's academics may blanche at the suggestion that they are influenced by religion or sentiment, but these "unscientific" ways of looking at animals have informed scientific thinking as well. Ecologists echo both the animal lover and the dominionist when they speak of our fellow creatures as units of broader categories—"native species," "the environment," "biodiversity"—that are valuable for their own sake, but which humanity is ultimately tasked with managing.

Their mindset also owes something to the contemporaneous development of modern science with industrialization, which offers the convenient, if disenchanting, metaphor of nature as an advanced machine to be tinkered with—and perhaps improved upon. As the steam engine powered technological advances and social upheaval in the 18th and 19th centuries, "[n]othing could escape the mechanical maw," wrote the environmental historian Donald Worster, in his 1977 classic *Nature's Economy: A History of Ecological Ideas*:

> In ecology the argument went that, like a planet in its orbit or a gear in its box, each species exists to perform some function in the grand apparatus . . . Man's conjoining species, perhaps himself as well, and the ecological order they formed would be studied as a lifeless, impersonal fabrication, admirable in its cleverness, perhaps, but not suitable for any emotional investment by men and women.

Even so, a romantic countercurrent continued to run through Western thought about nature, represented by poets like William Wordsworth, mystics like John Muir, and even scientists themselves, when they spoke of "ecological communities." As ecology and adjacent fields became professionalized in the 20th century, it would take a deliberate effort to finally purge this last vestige of sentimentalism. In 1935, the Oxford botanist Sir Arthur G. Tansley took a major step, offering the term "ecosystem" as an alternative to the concept of nature as a "community" or "complex organism," then-popular terms that he deemed "illegitimately derived and which introduce misleading implications."

This semantic adjustment was more consequential than it first appeared. Scientists had been using the prefix eco-, from the Greek *oikos*, or "household," at least since German naturalist Ernst Haeckel coined "ecology" in 1866 to denote "the science of the relations of living organisms to the external world." (Haeckel's term makes sense if one imagines organisms as members of a household who must manage resources and relationships with one another.) "System," from a Greek term meaning "a whole compounded of parts," anchored Tansley's new concept to a fully materialist worldview. Now, the study of nature involved not observing the *relationships* between plants, animals, and their environment but tracing flows of matter and energy through a *system* made up of biotic and abiotic components.

Tansley's ecology was a mechanistic and quantifiable science, in which "all living organisms may be regarded as machines transforming energy from one form to another." The ecosystems through which this energy flowed, meanwhile, represented "basic units of nature . . . one category of the multitudinous physical systems of the universe, which range from the universe as a whole down to the atom." (A patient and disciple of Sigmund Freud, Tansley attempted to conform human psychology to a similar model, in which the mind was "a very complex mechanism" powered by the flow of "psychic energy.")

By the time Charles Elton published *The Ecology of Invasions by Animals and Plants* in 1958, "ecosystem" had become a standard concept, one that is now central to the entire question of invasive species, which in a mechanistic model are like wrenches thrown in the works. But looking back from the perspective of the 2020s, Lynn argues that this "systems" way of thinking has helped desensitize mainstream ecology to violence against nonhuman animals—and that cultivating such an attitude has become a core function of education within ecology and related sciences.

"There's an ideological effort to marginalize animals as individuals, as thinking, feeling beings," Lynn says, and to see them instead as "biological machines or functional units of ecology or deliverers of ecosystem services."

Such an outlook perhaps made it possible, in the 1960s, for Edward O. Wilson and Daniel Simberloff to fumigate islets in the Florida Keys, summarily destroying whole populations of insects and spiders in the interest of better understanding them. Today it helps make palatable deadly campaigns against invasive cats or squirrels, conducted in the name of preserving native ecosystems. But suppressing qualms doesn't mean they disappear completely.

For many years, Lynn has organized workshops about conservation ethics for students, government personnel and the staffs of environmental nonprofits. He has noticed that many of the thousands of attendees suffer from a particular internal conflict.

"I have *never not* had people come to me in tears, telling me that moral desire to do right by other animals is what got them into the field," he says. "And they feel beaten down by their colleagues, by the requirements of their profession, and they can't say how they feel, because the professional standards are such that they will lose their jobs or lose opportunities if they do."

Looking up, Lynn is briefly distracted by a red-tailed hawk (*Buteo jamaicensis*), which has perched quite close to his office window, having chosen a sheltered spot in which to tear apart a songbird.

"This person is eating someone!" Lynn says. "Oh my, right there . . . " He pauses, rapt. "I enjoy watching it. It's gruesome, but it's amazing at the same time."

On the one hand, the hawk and the songbird are simply playing their assigned roles in the ecosystem of a deciduous woodland. But when the drama is this close, it becomes absurd to think of these two animals, magnificent predator and unfortunate prey, as merely a couple of machines transforming energy from one form to another.

Down the Rabbit Hole

Within academia, Lynn's is a countercultural position, but his side is growing. In recent years, a small, vocal group has argued that, as conservationists work to ensure the survival of biodiversity, native ecosystems, and endangered species, they must also account for the welfare of sentient animals, which—regardless of their place in the world—have moral worth as individuals. Since 2013, many of these scientists and philosophers have been associated with the Centre for Compassionate Conservation, or CfCC, housed at the Transdisciplinary School, a deliberately heterodox unit of the University of Technology, Sydney.

"It made sense to have it at UTS, because UTS was a very open-minded, still relatively young university," says Daniel Ramp, who founded CfCC in 2013. While the group has international affiliations, it also made sense that it would arise in Australia—not because the country is more compassionate than any other, but because its unique natural and political history has led to greater violence in the name of conservation than perhaps anywhere other than New Zealand.

"In Australia, colonialism has driven what I hesitantly, reluctantly, have to call hate," says Ramp, who grew up Down Under in the 1970s and '80s. "It is vile, the way we treat animals here."

Ramp originally trained as a botanist, but he became interested in questions of animal welfare after he began attending conservation conferences where "90 percent of the talks were about how to kill animals. And that struck me as odd." Questioning that, he soon realized, was a sure path to professional trouble. After Ramp founded his center there, UTS began receiving letters from other academics who called for it to be shut down. "They said that we were not being scientific, that we were animal rights activists dressed up as scientists—as if there's actually something wrong with that," Ramp says. "And those campaigns have continued."

While Ramp's critics might deny that they hate animals, managing wildlife with violence was common practice long before any Australian ever heard the phrase "native ecosystem." In the 19th century, European settlers and their descendants routinely slaughtered inconvenient creatures, including now-iconic species like kangaroos and emus, in the name of agriculture, sport, or fear (the Tasmanian tiger—*Thylacinus cynocephalus*—had a government bounty on its head almost until its extinction in the 1930s). Even in a more enlightened era, Australian society still condones bloodshed in the name of ecology: look no further than the industrial-scale culling of kangaroos in areas where they are deemed overabundant. Indeed, like their neighbors in New Zealand, most Aussies are far from timid about killing undesired animals, and they have long been at the technological vanguard when it comes to doing so.

"We have a national carrier TV station here called ABC [Australian Broadcasting Corporation], and every week they have a gardening show where they'll use the word 'native' every minute," Ramp says. "There's a complicit-ness across society that I think is born out through colonization."

As in New Zealand, it's difficult to untangle Australia's present-day posture toward non-native species from its history as a settler-colonial enterprise, which involved not only importing countless species from Europe—either deliberately or by accident—but also the alteration of the continent's landscape to better suit European tastes and economic pursuits. The shift toward persecuting non-native species is in a way a continuation of that history—as well as, perhaps, an effort to displace blame.

"The historical foundation of Australia was the eradication of our Indigenous people, of indigenous fauna and flora," Ramp explains. "There is this really deep-seated legacy that's ever-present. And there's quite a myopia around history."

European settlement of Australia kicked off in 1788, when eleven British ships, carrying the first few hundred of what would eventually number 150,000 convicts, arrived in New

South Wales. Despite its origins as a penal colony, within a few decades the settlement was prosperous enough to boast an aspiring upper class, which yearned to enjoy the traditional pursuits of the English aristocracy—namely, shooting at small game in a bucolic setting. In 1859, one such wealthy colonist, Thomas Austin, received a shipment of twenty-four rabbits (*Oryctolagus cuniculus*) from a relative in England and, with the intention of organizing hunting parties, released them on his 29,000-acre sheep farm in Victoria.

Finding themselves in a landscape that may have felt like home, the English rabbits took full advantage of their freedom and set about breeding. They were massively successful: six years later, hunters shot forty-four thousand on Austin's estate alone—and that massacre barely made a dent in their population. By 1900, *O. cuniculus* ranged across most of the continent, making a laughingstock of human efforts to crisscross the landscape with so-called "rabbit-proof" fencing.

During the Great Depression, Australians enjoyed the abundance of free, easy-to-catch meat. Still, by the late 1940s, there were an estimated six hundred million rabbits—eighty for every Australian man, woman, and child, who, though many did their part, couldn't hope to eat through such abundance. Nor could they trap, shoot, run over with heavy machinery, or poison enough rabbits to bring down their numbers. But this was an age of innovation, and as farmers complained of a plague of rabbits eating their crops and digging up their pastures, Australia turned to biological warfare.

In the 1950s, the national science agency CSIRO (Commonwealth Scientific and Industrial Research Organisation) introduced the myxoma virus, which originated in South America. Spread through close contact as well as by mosquitos and other biting insects, the virus is harmless to all but rabbits and hares, in which it causes a disease called myxomatosis. Symptoms include ulcers in the mucus membranes, fever, and blindness. Swelling around the mouth and nose leads to difficulty breathing, and infected animals usually die within two days.

In Florida, using insects to manage non-native plants like Old World climbing fern feels like gently restoring the balance of nature. Australia's rabbit program on the other hand, took the concept of biological control to what was arguably a darker place. It was also more effective than the efforts of the USDA have ever been in the Everglades—at least at first. Within a few years, the myxoma virus had killed more than 99 percent of the continent's rabbits, a feat that remains one of CSIRO's proudest achievements.

"The change . . . has been almost miraculous: the landscape in some areas has been virtually transfigured," wrote CSIRO scientist F.N. Ratcliffe in 1956:

> Hills that had been grazed to the soil for decades, and whose slopes had appeared grey and red on the horizon, are now clothed in grass. The broad margins of country roads, lying outside the boundary fences of grazing properties, tended to carry dense rabbit populations and as often as not showed it in the poverty of their ground cover. It is now usual to see tall grass and herbage to the road's edge. Looking over the fences, it is now very rare to see a paddock without a dense and healthy pasture.

Perhaps more to the point, sheep farming—then Australia's largest industry—was made more profitable by the elimination of leporine competition. According to a 2013 estimate, "biological control of rabbits produced a benefit of A$70 billion" for

Australian agriculture in the sixty years after myxomatosis was first introduced.

This glorious change was not fated to be permanent, however. Among the few rabbits who survived, some passed resistance to the virus on to their young, and by 1995 the rabbit population had rebounded to an estimated three hundred million. CSIRO was forced to seek new weapons. In the 1990s, it introduced calicivirus, a pathogen discovered in the 1980s in China that causes rabbit hemorrhagic disease (the hallmark symptom is sudden death). But rabbits developed resistance to that virus, too, forcing CSIRO to hunt for newer, more lethal strains. Today, according to the NGO Rabbit-Free Australia—motto, "Bilbies Not Bunnies"—the country still has some two hundred million rabbits, down from what the group presumes is a peak of ten billion (a historical figure that, to be clear, dramatically exceeds other estimates).

Rabbits are not the only public enemy to face the lethal power of Australian ingenuity. Recently, the Centre for Invasive Species Solutions (CISS) at the University of Canberra rolled out PAPPutty™, a lick-able paste containing the poison para-aminopropiophenone, "developed to enable the self-euthanasia of wild dogs and foxes captured in foothold traps." And in 2023, authorities in the state of Western Australia green-lighted the Felixer™, a solar-powered device, equipped with sensors powered by artificial intelligence, which can identify passing cats with a high degree of accuracy and spray their fur with toxic gel. When the cats groom themselves, as all cats do, they will ingest the poison and . . . self-euthanize, as the folks behind PAPPutty might put it.

The Felixer has yet to be widely deployed, but conservation officials have high hopes that it will help reduce the population of a species deemed responsible for dispatching twenty-seven of the thirty-four Australian mammals that have gone extinct since Europeans arrived. Such evocatively named natives as the pig-footed bandicoot, the Nullarbor dwarf bettong and the broad-faced potaroo are now lost to history, but, echoing Peter Marra's research in North America, the CISS has estimated that feral cats and roaming pets continue to eat some two billion mammals, birds, reptiles, and amphibians each year, plus more than a billion invertebrates, including valued native grasshoppers, beetles, and butterflies. Cats are, according to one ABC documentary, "native animal annihilators."

While killing them seems a straightforward solution, even those who hope to eliminate feral cats altogether acknowledge that poison is at best a necessary evil. Despite being billed as "humane," the poison used by the Felixer, sodium fluoroacetate (also known as "compound 1080") has a grim history. Odorless, tasteless, and colorless, it was first patented in Germany around 1930; the Nazis reportedly considered using it in chemical warfare against Allied troops, or to kill prisoners in their death camps, before deciding it was too dangerous. Indeed, the chemical is so toxic that by the late 20th century it had been banned in most of the world, with the notable exceptions of New Zealand and Australia.

Compound 1080 causes protracted suffering, including nausea, vomiting, abdominal pain, difficulty breathing, convulsions, and paralysis. But are the slow, excruciating deaths of feral cats redeemed by the prospect of saving the native fauna they might have eaten? Greater bilbies and burrowing bettongs—or at least, those who advocate for these native marsupials over non-native predators—may say yes.

Arian Wallach, for one, is a hard no.

A native of Israel, Wallach moved to Australia to study dingoes, earning a doctorate at the University of Adelaide and then spending nearly a decade at the Centre for Compassionate Conservation. In 2022, she joined the faculty at the Queensland

University of Technology in Brisbane. She now lives half a day's drive outside the Queensland capital on five acres of bushland, a not-quite-ranch with a less-than-reliable internet connection. Wallach is not an easy person to reach, and when we finally connect over Zoom she apologizes for being late, explaining that she "had some cow issues."

Historically, keepers of livestock have not been overly fond of predators, but Wallach is a proponent of coexistence, even with predators of dubious native status—for instance, the dingo, which likely traveled to Australia from Asia between five thousand and twelve thousand years ago in the company of human sailors. Wallach's tolerance also extends to more recent arrivals, including cats and foxes, and she disputes the case that Australian conservation groups and land managers have made against them.

"If you have a big project to kill cats, you get a pile of dead cats," Wallach says. "But do we *know* that cats have caused extinctions?"

Extinction is rarely monocausal, and with the exception of certain marquee cases on islands—mostly flightless birds, the proverbial sitting ducks—the case that predators are responsible for dispatching any particular creature is rarely so cut and dried.

"We can think about it like a murder case," Wallach says. "Who was there at the time, who has the means, who has the desire? You might find that the mammals attributed to cats and foxes went extinct after cats and foxes arrived, maybe within a certain timeframe. That's circumstantial evidence. And in a large proportion of cases, there's not even evidence of that."

In many cases, Wallach notes, the Australian animals whose disappearance is attributed to introduced predators are themselves quasi-mythical, known from a handful of bones, a sketch, or reported sightings decades or centuries ago. In those cases, "how do you then deduce what happened?"

In arguing for the defense, Wallach challenges a longstanding contention that introduced species are the second greatest threat to biodiversity worldwide, after habitat destruction. That oft-repeated claim—according to a 2015 forensic investigation by Matthew Chew—originated with Edward O. Wilson's popular 1992 book *The Diversity of Life*, in which the celebrity biologist extrapolated the idea from a back-of-the-envelope calculation based on extinctions of North American freshwater fish. Through a sort of peer-reviewed game of telephone, the assertion would soon take on the mantel of "scientific consensus."

Wallach has long doubted this received wisdom, but it wasn't until she was working on a paper with her former student Erick Lundgren that she glimpsed the true fragility of the claim's foundation. The study, published in 2025 in the journal *BioScience*, reviewed evidence that cats and red foxes (*Vulpes vulpes*) were responsible for more than half of extinctions (or near extinctions) of Australia's endemic mammals.

"I dug up everything I could possibly find," she says. "And I mean, every anecdote, every instance of, 'Oh, I saw a cat with a whatever in their mouth,' or anything. For a large percentage of these species that have been attributed to cats and foxes, there's nothing. There's just a whole pile of nothing."

Wallach and Lungren looked at more than 160 population crashes, involving some fifty species, that Australian experts and officials have attributed to fox and cat predation. In most cases, they found no evidence even of a correlation with either predator's arrival in an area, let alone a direct causal relationship. Some native animals may have actually disappeared before these predators arrived. Others seem to have coexisted with cats or foxes—the latter, like rabbits, were also introduced by would-be gentleman hunters in the mid-1800s—for a century or more before the population began to decline. Others are obviously too big for introduced predators to be their biggest problem, despite conservation officials' assertions—for example, the rare yellow-footed rock-wallaby (*Petrogale xanthopus*), which can clock in at more than eighteen pounds, more than twice the weight of the average tabby and significantly more than most foxes.

Indeed, Wallach and Lundgren wrote, the "self-evident truism" that cats and foxes were to blame was quite possibly neither evident nor true: "Expert opinion can be useful, but it is prone to bias, groupthink, and overconfidence . . . and it should not be confused with evidence." The study did not, they were careful to note, fully exonerate these introduced predators. But perhaps there was not quite enough proof of guilt to justify the death penalty.

Animal Lovers

Teasing out the proximate causes of historical extinctions is more than an intellectual exercise. It also has implications for the cats, foxes, and other non-native predators that—for the moment, at least—are alive and stalking the landscape today. But if we consider introduced species innocent until proven guilty, we may also be complicit in the loss of other species that might one day be eaten to extinction, infected with novel diseases, or outcompeted in an altered environment.

Extinction has been a risk throughout the history of life on Earth, but it has become more common as human activity speeds up the pace of change, creating a sense of urgency among those who hope to hold on to the species we still have. That's why those Italian scientists took the grey squirrel matter into their own hands, and it's why Peter Marra has no patience for fellow scientists who question the need to do something about outdoor cats.

"I think they're causing harm," Marra says. "I mean, everything dies at some point, and it's not 'humane.' Cats in particular—they get hit by cars—I mean, the data are really pretty clear. Cats outdoors don't do well. To just leave them out there . . . what is compassion? Who defines compassion? What about the birds that they're killing?"

He also despairs of ever reaching the people who feed, look after, and campaign for feral cats.

"I don't think, with the cat advocates, we've had any impact," Marra says. "I really value their passion for animals—I'm passionate about animals. I'm passionate about nature. I've been very fortunate to have had remarkable experiences, and people that don't have access to nature in those ways, I feel like cats are almost a substitute. So I understand how cat advocates feel about this horrible person, Pete Marra, who feels like we should be euthanizing cats. But at the same time, I think about the whole ecosystem, and about what cats are doing and how we have to make challenging decisions immediately."

Peter Wolf is one of those cat advocates who will never come around to Marra's point of view. His posts at *Vox Felina* have grown more sporadic in recent years, but while the fury over *Cat Wars* may have died down, Marra's old adversary remains passionate about the plight of unowned felines, and he's still poised to attack anyone who threatens their lives—or even their self-determination.

Wolf shares his condo in central Phoenix with five rescued pet cats, but he cares for many more that would never make it on the inside, including a couple dozen neighborhood moggies that he feeds and checks up on, and the handful of regulars that visit his patio for a nightly "buffet." His longest-running customer is an elderly female he calls "the Mayor."

"She still patrols the place and still gives me the most withering look," Wolf says affectionately. "I can't touch her. All these years she's never missed a meal, and yet she still seems to hold a grudge."

The Mayor's attitude likely stems from innate temperament rather than any hard feelings, but Wolf does admit to trapping her and the litter she brought along soon after they appeared at his complex more than a decade ago. He found homes for the kittens, but the Mayor seemed determined to maintain her independence. So, after Wolf had her spayed and vaccinated, he freed her to go about her business.

"There's nothing I can do to protect her out there," he says, shaking his head. "There's a kind of constant worry that comes along with it." He wishes he could tell her, "Look, if you want to retire indoors, all you got to do is come by, knock on the window." But it's more likely she'll simply stop showing up one day, and he'll never know what happened.

It's frustrating to care about animals you can't really protect, but Wolf, who has loved cats since he was a boy, is committed to doing what he can for them—both those under his direct care and felines in general. Though he had a career as a mechanical engineer and industrial designer, in recent years protecting cats has become his life's work. In 2013, he joined the staff of Best Friends Animal Society, a nonprofit dedicated to promoting pet adoption and no-kill animal shelters.

"I thought, well, this would be great because instead of having a day job that was completely disconnected from animal welfare, now there's some built-in efficiencies," he says. "Instead, it was like, okay, do you want to spend your day on what you're already spending all your evenings and weekends doing?"

While they are well-known to one another, the two Peters have never spoken, and indeed both seem reluctant to ever do so. Marra remains gun-shy, and despite his expressed wish for common ground, he believes cat advocates are "too far gone" to see his point of view. Wolf, for his part, maintains that Marra is a "post-truth pioneer" in the style of Donald Trump. But despite their mutual distrust, they have much more in common than their first names.

I'm certain, I tell Wolf, that both he and Marra have voted for the same presidential candidates—and I don't mean Trump. I bet these two well-educated, middle-aged white men get their news from the same sources, and I'm sure they have similar opinions about nearly every other environmental issue, from habitat loss to climate change. Is it simply the narcissism of small differences that makes this particular topic so heated?

"We probably do share a lot," Wolf acknowledges. "Looking at my bookshelf, yeah, we probably actually do have a lot of the same books and probably consume a lot of the same media . . . I guess, yeah, I mean, the idea that we care a lot about animals . . . I had never thought about it, but I'm quite sure you're right."

Both men also have an uncommon willingness to speak out on issues that matter to them, with the same sort of fire. And they share a commitment to fights that may end only in uneasy draws.

Trap-neuter-return, for example, is a long game, one that aims to decrease the population of feral cats through gradual attrition. Each side in the cat wars can point to studies that show TNR does or doesn't work, but like the question of how bad cats are for birds, this isn't an argument that can be won by even the best-designed study or most thorough meta-analysis. Instead, it's a conflict of values. Do individual cats have a right to pursue their lives as they see fit, with humans simply doing our best to make life easier for all parties? Or are outdoor cats—which inevitably kill some number of other animals—an unnatural scourge on the natural world that humans are obligated to remove?

Wolf has plenty of arguments for the practice, but day-to-day, or night-to-night, TNR is also a burdensome commitment, and not dissimilar to the work of humans who place themselves on the front lines of the war on invasive species. Like a squirrel-free park in Italy or a patch of Michigan forest that's been cleared of garlic mustard, a well-cared-for colony of neutered cats is only stable until the next irresponsible person drops off a box of unwanted kittens.

"It can be terribly frustrating, because it seems like so much work, and it sort of seems to never end," Wolf says. "Say, 'Okay, we've got things under control,' but then trouble creeps in again."

There are parts of Phoenix where, a decade ago, car headlights would illuminate the eyes of countless cats, Wolf says. "You'd just pick up eyeballs everywhere—'Oh my God, how many cats?'—and that's no longer the case. So, it is satisfying. But again, if you turn your back?"

During the COVID-19 pandemic, veterinarians closed their doors, canceling appointments to neuter cats. "Then there was a bumper crop all of a sudden: 'What the heck are we supposed to do with all these kids?'" Marra says. "That gave us a little insight into, oh boy, this is ongoing! You got to be vigilant."

9

BRAVE NEW WORLD

For years, the deer problem was one of Ann Arbor's most divisive issues.

White-tailed deer are native to North America, but if they had come from anywhere else they would likely top the list of America's worst invasive species. Deer devour flower gardens, young oak trees, and spring ephemerals—and they seem to prefer the rarest and most beautiful plants, leaving less desirable species (like garlic mustard) to grow unmolested. Healthy adult deer have no natural predators, and like many an invasive species they thrive in the fractured suburban habitat that now characterizes so much of the modern United States. Seemingly unafraid of humans, they are highly visible in cities and towns, and their inability to understand traffic creates a perennial hazard for motorists.

But deer are also beautiful and beloved, icons of American wilderness that people still thrill to see, no matter how many hostas they eat or highway crashes they cause. And in 2015, when Ann Arbor announced a program to cull them within city limits using sharpshooters, many residents were outraged. Community groups with names like FAAWN ("Friends of Ann Arbor Wildlife in Nature") organized in opposition, and "Stop the Shoot" yard signs popped up everywhere. Even as annual deer culls continued from 2016 to 2020, however, the city offered a compromise to animal welfare advocates. Around one hundred deer each year would be killed, while another eighty does would merely be captured and spayed.

"I feel it is inhumane to kill animals who are completely innocent," Lorraine Shapiro, president of the group Ann Arbor Non-Lethal Deer Management, told a local newspaper. But although the group still opposed killing deer, she said, "[w]e decided to take a pragmatic approach. We can save a bunch of deer through sterilization."

As in the trap-neuter-return strategy for controlling feral cats, humane treatment was the program's top priority. After darting does with tranquilizers, city contractors blindfolded and transported them on stretchers to a field hospital set up at a local golf course, where veterinarians removed their ovaries under general anesthesia. The animals recovered from their operations on yoga mats before returning to the parks and other green areas of the city.

According to a report by the Michigan Department of Natural Resources, which gave Ann Arbor special permission to sterilize deer under the auspices of a research project, between 2016 and 2020 this combination of culling and fertility control helped cut deer populations roughly in half across the city. In addition to no longer being able to reproduce, the spayed females did not undergo a normal hormonal cycle, didn't go into heat, and thus didn't attract bucks from far away, as fertile does usually will. That meant fewer deer crossing roads and less conflict with humans.

It wasn't a perfect solution. For one thing, sterilizing deer costs a lot of money—nearly $1,600 for each doe, almost four times the per-deer cost of hiring professional hunters. And just like trap-neuter-return, sterilizing deer only works as long as people continue to do it. The COVID-19 pandemic put an end to Ann Arbor's deer control project, and while the white-tailed population has yet to rebound to 2015 levels, from time to time city officials propose another cull. This very afternoon, I watched a doe and her twin fawns amble down a residential street in the southeast section of town.

It seems that it's only a matter of time.

Still, fertility control remains an alluring proposition as a way of dealing with undesirable animals without killing them. And scientists may soon have a way to do it that doesn't require continuous upkeep. It's a set-it-and-forget-it method that would make use of evolutionary principles to reduce animal populations without killing a single creature.

Mother's Curse

Fertility control has long been used to curb the populations of undesirable organisms—not only via surgical sterilization, as with feral cats or Ann Arbor's deer, but also with chemical contraception, distributed via bait stations to animals like wild boars (*Sus scrofa*). Perhaps most successful, although so far only in arthropods, is the "sterile insect technique," which was developed in the mid-20th century to combat agricultural pests and disease vectors. In the classic version of this procedure, technicians rear insects in a laboratory, separate the males, and irradiate them, rendering them infertile but otherwise intact. The sterile males are then released to mate with wild females, who go on to lay eggs that will never hatch, and gradually the overall population dwindles. The technique has been used in mosquitos and fruit flies, and it enabled the elimination of the New World screwworm (*Cochliomyia hominivorax*), a deadly—but native—livestock parasite, from North and Central America by the 1990s.

These approaches, however, are all fairly labor-intensive, and they only work as long as people continue doing them. Long after rearing facilities that once churned out sterile screwworm flies by the tens of millions fell into disuse, the species popped up again north of Panama; today, screwworms have recolonized Mexico, and US authorities fear it won't be long before they reach the vast cattle ranches of Texas.

But what if nature could do the job on its own? In the 2010s, a group of scientists in New Zealand came up with a theoretical way to produce sterile males without irradiation, birth control bait, or even surgery. It would involve a basic quirk of sexual reproduction.

"The idea is that you could harness male-sterilizing mutations in the mitochondrial DNA," explains Damian K. Dowling, head of the Experimental Evolutionary Biology Research Group at Monash University, in Melbourne, Australia. "If it worked, it would be, economically, much more viable than the current approaches."

Often referred to as the "powerhouses" of cells, mitochondria are found in almost all eukaryotic organisms, from amoebas to elephants. They first appeared at least 1.45 billion years ago as endosymbionts—that is, single-celled organisms that came to live inside larger host cells, contributing to their hosts' fitness by processing oxygen and glucose into usable cellular fuel. A legacy of this origin is that they retained their own genomes, DNA separate from the nuclear chromosomes that carry the blueprint for building all the other components of an organism. And in nearly all plants and animals that reproduce sexually, only the female gamete—the egg—contributes mitochondrial DNA to the offspring (mitochondria present in the sperm cell dissolve as it delivers its payload of paternal chromosomes). That means that, with certain rare exceptions, every organism's mitochondria are essentially identical to its mother's. The pressure that natural selection exerts on these genes is thus only felt by females, since for mitochondria males represent an evolutionary dead end.

"If these genes are not negatively affecting female fitness and reproduction, then they can be passed on to the next generation, and the daughters will pass them on to their offspring," says Neil Gemmell, a geneticist at New Zealand's University of Otago, who has been studying the effects of mitochondrial DNA on viability and fecundity for decades.

It was Gemmell who, in 2013, proposed harnessing this phenomenon to control pest animals. He called it the Trojan Female Technique, for the mythical wooden horse that the people of Troy brought through their own gates, unaware that a Greek raiding party hidden in the animal's belly would soon emerge to destroy their ancient city. Like the crafty Greeks, mitochondria with genes that harm males—for example, by hampering the motility of their sperm—could be smuggled into a population by unaffected females. Over time, there would be fewer and fewer fertile males in each generation, leading to a gradual decline in the population as adults die off and fewer offspring take their place. No poison, no traps, no guns required.

Modeling suggests that a single large release of females with mutated mitochondria, amounting to about one-tenth of the target population (or several smaller releases of about one one-hundredth) could suppress a population to near zero within decades, depending on how frequently the target species reproduces. Meanwhile, Dowling's lab has tested the effects of male-harming mitochondrial genes in the common fruit fly (*Drosophila melanogaster*) and found that the Trojan Female Technique could quickly suppress caged populations.

Promising though it may sound, there are still some hurdles science must overcome before the Trojan Female can be used in the real world, Dowling cautions. For one thing, experiments so far have required naturally occurring mutations. While editing nuclear DNA has become almost routine thanks to the technique known popularly as CRISPR, making deliberate changes to mitochondrial DNA is a bit trickier because the short segments of RNA that guide the CRISPR "scissors" can't get through the double membranes of these organelles.

"It's been really hard to develop those same sorts of gene editing technologies for mitochondria," Dowling says. "When we can do that, then in theory, this technique will be a lot closer to market than it currently is."

In the past few years, scientists at the Broad Institute of MIT and Harvard, in Cambridge, Massachusetts, have made strides in mitochondrial DNA editing, with the ultimate goal of treating human mitochondrial diseases. But while these early-stage developments also bring the Trojan Female closer to feasibility, there remains the question of whether genetic approaches to pest control can ever gain public acceptance.

In New Zealand, a smattering of focus groups and surveys looking at how the public might feel about the Trojan Female Technique suggests Kiwis, at least, are interested—but also wary.

"I think it's got more potential than anything else, but there is a saying that says that, if something can go wrong, it probably will go wrong," as one focus group participant put it.

Others, perhaps inevitably, brought up both Dr. Frankenstein and the idea of "playing God," comparisons that have long dogged scientists who seek to modify genomes, especially for the purpose of eliminating undesired organisms.

Austin Burt, an evolutionary geneticist at Imperial College London, first conceptualized the synthetic gene drive in 2003. Similar to the Trojan Female Technique but with broader applications, it involves creating a gene variant with a greater-than-usual chance of being inherited by the next generation. Scientists might, for example, add code for proteins that would then change wild-type variants of the gene to match the engineered version, ensuring that an altered organism's offspring will always carry two copies—and the wild type will eventually disappear.

Gene drives could enable disease resistance, in the case of crops, or they could reduce an insect's ability to transmit a disease to humans. In invasive species applications, the goal would be to reduce the organism's fitness or fertility, eventually eliminating the population by natural attrition. But while many labs have published proof-of-concept studies, and public health and conservation NGOs have seriously explored gene drives as solutions for unwelcome creatures like island-dwelling rodents and malarial mosquitos, a so-called social license remains elusive.

Gemmell, for example, harbors skepticism of engineered gene drives.

"First thing I asked was, how do you turn them off?" he says. "It still hasn't been answered." While some have argued that another gene drive could end the first, he says, that "sort of felt to me like trying to clean up uranium with plutonium."

In 2017, Gemmell teamed up with MIT biologist Kevin Esvelt to highlight the potential pitfalls for a commentary in the journal *PLOS Biology*. They warned that igniting a self-propagating gene drive might be "equivalent to creating a new, highly invasive species—both will likely spread to any ecosystem in which they are viable, possibly causing ecological change. . . . Even assuming that national sovereignty is morally irrelevant, the social and diplomatic consequences of an unconstrained release should give us pause."

It was a head-turning assertion, especially coming from Esvelt, who was the first scientist to propose using CRISPR for self-propagating gene drives, in 2013. But the paper stirred up particular controversy in New Zealand, where policy-makers were growing increasingly enthusiastic about what CRISPR-based gene drives might do for their invasive species problems.

"There was this idea that gene drives were going to be great, a be-all and end-all for pest control in New Zealand," Gemmell says. "We wrote a paper that didn't say gene drives were a bad thing. We just said, 'Hey, you need to be a bit careful about how you use 'em.'"

What Gemmell still considers to have been a fairly measured warning proved more controversial that he realized. "We don't have Senate inquiries like you have in the States, but I got pretty close to that level of government reaction, where they sort of hauled me in and said, 'What the hell are you saying?' and, 'You're doing damage to the country.' And I was like, 'No, I'm not. We are putting forward a view. It's not saying that New Zealand's making a mistake. This could be great. We just need to be careful.'"

In short, he says, "I made no friends for myself."

Reversing the Trojan Female might be more straightforward than switching off an engineered gene drive, Gemmell says, arguing that one could simply breed wild-type females and release them back into the population until the prevalence of the male-harming mutation was reduced. Still, he and other proponents acknowledge that even this less drastic scheme, should it become technically feasible, ought not to be embarked upon lightly.

"I think there are really legitimate questions before this rolls out," Dowling says. "How can we ensure that there isn't any inadvertent migration of our pest populations back to their native ranges? And how can we mitigate the risks of that leading to some catastrophic outcome?"

Still, he adds, authorities in Australia have, for years, controlled the rabbit population with deadly viruses, and so far those pathogens haven't had any unintended effects beyond rabbits' pesky tendency to evolve resistance. And, as many have noted, humans have in effect been playing God for centuries as we transport species from one place to another—without considering whether God, or perhaps evolution, might have meant for them to stay put.

Practical and spiritual qualms aside, however, there's a certain comfort to the idea that bleeding-edge 21st-century technology might make moot all our worries about invasive species, without any spilled blood. After all, our ingenuity has gotten us out of tricky situations before: Antibiotics, vaccines, sanitation, and the Green Revolution in agriculture have all made problems that once seemed intractable vanish in short order. (Of course, some of these solutions have brought new problems, including antibiotic resistance and polluting agricultural runoff, but to the optimist these are also surmountable).

Using geoengineering to slow climate change carries a similar allure—indeed, it is so tantalizing that in 2025 the United Kingdom's Advanced Research and Invention Agency announced a £50 million government-funded program that will sponsor outdoor experiments to block the sun's rays by sending aerosols into the stratosphere.

However, even as optimists pour money into researching quick fixes (and as some of us keep doing our best to reduce greenhouse gas emissions), in recent years more and more scientists, governmental bodies, and international organizations have shifted their messaging about climate change to include the need for humanity to adapt to a warming planet. That may mean concentrating resources on the most vulnerable people and ecosystems, shoring up infrastructure, rethinking everything from city planning to agriculture, and restoring wetlands to protect coastlines from rising sea levels. We will also have to recognize what we can fix, what we can't, and when we might just have to accept that our world has changed forever.

And that may be just as necessary when it comes to dealing with non-native species.

Loving the Alien

First, the necessary caveats: the mistakes of previous eras teach us that recklessly spreading organisms from one continent to another may have unpredictable consequences, and—just as we really should stop building coal plants—we would do well to minimize new introductions, whether deliberate or accidental. Meanwhile, some existing invasive species are clearly causing harm, at least from the perspective of human societies. Even if they can't be eliminated, controlling their numbers is prudent.

Sea lamprey (*Petromyzon marinus*), a cartilaginous fish, spread to all of North America's Great Lakes via the St. Lawrence Seaway soon after the Welland Canal connected Lakes Ontario and Erie in 1829. Lamprey parasitize bony fish, latching onto their flanks with sucker mouths and feeding on blood and fluids until their hosts weaken and die. This marine species, which has few freshwater predators and is at best a niche human foodstuff, helped deplete populations of trout, whitefish, and sturgeon—fish that are not only keystone species in the Great Lakes food web but also commercially important across the region.

Only the most resolute compassionate conservationist would argue that the United States and Canada shouldn't work together to control sea lamprey—even if the traps, barriers, and lampricide poison deployed in the effort might be considered cruel from a lamprey's point of view. While these measures will never eliminate the species entirely, they are narrowly targeted and apparently effective. According to the international Great Lakes Fishery Commission, the lamprey population is down 90 percent from its height. Meanwhile, lake trout, which a combination of lamprey predation and overfishing once brought close to extinction, are increasing their numbers. In 2024, they were declared fully recovered in Lake Superior. Maintaining that success costs $25 million annually, but the Great Lakes fishery is a $7 billion industry that supports seventy-five thousand jobs: economically, at least, the math checks out.

Other cases are less clear-cut, even those that appear superficially similar. While the United States and Canada agree that lamprey are *pesca non grata*, halfway across the world there *is* a constituency for Nile perch (*Lates niloticus*), a large carnivore first introduced into East Africa's Great Lakes under somewhat mysterious circumstances in the 1950s. As laid out in the 1996 book *Darwin's Dreampond* by the Dutch biologist

Tijs Goldschmidt, once they arrived in Lake Victoria the 450-pound perch began devouring the native cichlids (genus *Haplochromis*). Like their counterparts in Lake Malawi, these colorful fish are among the most evolutionarily interesting groups of closely related species in the world. But ecologists who study them now blame perch at least in part for the extinction of two hundred of the lake's five hundred endemic cichlids; they call the situation an ecological catastrophe.

The thirty million Ugandans, Kenyans, and Tanzanians living around the lake, however, see Nile perch quite differently. Between the 1970s and 2000s, their annual catch of fish grew from thirty thousand to five hundred thousand tons, with exports rising from $1 million to $260 million. This economic boom was due entirely to the sheer mass of a Nile perch, compared to the fairly paltry size of the cichlids that used to dominate Lake Victoria. There were new jobs for fishermen, fish processors, and wholesalers, along with all the other services that support them, for better or worse (that is, not just microfinance but also prostitution).

"Walking into a lakeside fishing community today and asking people how they feel about Nile perch is roughly analogous to walking into a General Motors plant and asking workers and job applicants how they feel about cars," wrote biologist Robert M. Pringle in the entry on Nile perch for the 2011 *Encyclopedia of Biological Invasions*. "They may or may not personally use the product that they produce, but their livelihoods depend on the global demand for it, and most are too young to remember the horse-and-buggy days."

As with Detroit's auto industry, however, the 21st century has seen a decline of the perch fishery, to the extent that African authorities now worry that perch may disappear from Lake Victoria altogether. That fear is exacerbated by a growing Chinese demand for fish maw, a delicacy made from swim bladders, which has increased the value of Nile perch, along with the lengths fishermen will go to in order to catch them.

There's some good news, though: as perch decline, scientists report that cichlid populations are bouncing back.

Some invasions upend entire ecosystems, but others turn out to be more smoke than fire. Kudzu, which David Fairchild so regretted planting in his own yard, was once known as "the vine that ate the South." But nearly a century after a major federal campaign to encourage the planting of kudzu as erosion control, it seems to have colonized only about 7.4 million acres, an area a little smaller than the state of Maryland. Whether or not this widely cited number is accurate, it's clear that only a small fraction of the invasion includes high-quality habitat: The US Forest Service has estimated kudzu can be found in only 0.1 percent of the South's two hundred million acres of forest. But kudzu is more successful as a colonizer of fallow fields and the disturbed land along roadsides—the parts of the South that people *see*. For that reason, the vine has become a popular metaphor.

Writing in *Smithsonian* in 2015, Alabama naturalist Bill Finch observed that for many writers in the mid-20th century,

> kudzu served as a shorthand for describing the Southern landscape and experience, a ready way of identifying the place, the writer, the effort as genuinely Southern. . . . But for others, kudzu was a vine with a story to tell, symbolic of a strange hopelessness that had crept across the landscape, a lush and intemperate tangle the South would never escape."

Kudzu was racism, poverty, and the weight of an inescapable past—or else, it was the indomitable Southern spirit, a rich culture that would triumph over any obstacle. In any case, the story of *P. montana* proved to be as much about people as it was about nature. That's true, of course, for plenty of invasive species—and meanwhile the natural world remains oblivious to the narratives we impose.

Like kudzu, tamarisk (genus *Tamarix*) was once championed by the US government. A brushy, flowering shrub or small tree that thrives in arid places around the Mediterranean, the Middle East, Africa, and Asia, tamarisk tolerates drought and salt, and in the Torah it symbolizes God's eternal covenant with Abraham and the endurance of faith.

According to an exhaustively researched historical account by Matthew Chew, the Americans first imported tamarisk as an ornamental sometime before 1818, but by the late 19th century the Army Corps of Engineers had pressed it into more practical service stabilizing the banks of shipping channels. Later, Mark Carleton, an employee of Fairchild's Office of Foreign Seed and Plant Introduction, promoted tamarisk for soil conservation on dry lands in the American West, writing in 1914 that it was "the most drought resistant and otherwise hardy of all the trees and shrubs" he had planted at his Texas ranch: "There appears to be no limit in dryness of the soil on any usual Great Plains farm beyond which this plant will not survive."

Like many of the USDA's early 20th-century darlings, by the middle of the century tamarisk—known colloquially as saltcedar—had both escaped captivity and fallen out of favor. As Chew puts it, "Shrubs once extolled for erosion and sedimentation control became machine-like monsters pumping away scarce western water." Soon state and federal authorities were attacking tamarisk wherever it grew, using poison, saws, and an aesthetically shocking piece of heavy machinery known as the LeTourneau Tree Crusher. (The US Army also used the latter to clear the jungles of Vietnam, as a sort of mechanical Agent Orange.)

But in the 1990s, new information complicated the picture of tamarisk as a botanical water thief without redeeming qualities. It turned out that the endangered southwestern willow flycatcher (*Empidonax traillii extimus*) had begun nesting in tamarisk's dense, sheltering branches.

The flycatcher, it must be noted, was endangered because humans had degraded its habitat. Development in the American West required damming streams, diverting water to farms, and pumping groundwater out of aquifers, all of which added up to the loss of a diverse, native riparian habitat that had once held prime nesting sites in cottonwood and willow trees. The brown-headed cowbird (*Molothrus ater*), a native brood parasite that lays its eggs in flycatcher nests, also became more abundant as people converted more acreage to pasture and cropland.

But these perky little songbirds found refuge in tamarisk, which flourished along the new, drier western streambanks. The southwestern willow flycatcher now depends so thoroughly on the invasive bush that after the USDA introduced the tamarisk leaf beetle (*Diorhabda carinulata*) as a biocontrol in 1999, flycatcher nests began to fail. The beetles were defoliating tamarisk during the height of nesting season, exposing chicks to heat and predation. For a time, things looked grim, with the subspecies down to about one thousand pairs in the early 2000s. USDA officials made an about-face, discontinuing permits for landowners to release the beetle and warning that any unpermitted transport risked violating federal law, with a potential $60,000 fine.

"Help us spread the word to stop the beetle's spread," pleaded an official communiqué, just a few years after the Agricultural Research Service had called the bug "an ideal biological control agent."

Another USDA import, the Chinese tallow tree (*Triadica sebifera*), has also eluded simple classification into botanical "good" or "evil." Sown along the Gulf Coast, the hardwood spread out of control through native prairies and marshes, apparently another regrettable mistake. Except that, like tamarisk, tallow trees also turned out to have an upside, at least for one particular constituency.

Now known popularly as the "popcorn tree," *T. sebifera*'s clusters of white flowers are among the first to bloom in the spring—as early as February—and the longest-lasting (through May). That makes them an important source of nectar and pollen for honeybees, which, although also not native, are valuable to humans as both producers of honey and pollinators of crops. Beekeeping trade groups, including the Louisiana and Texas Beekeepers Associations, the American Beekeeping Federation, and the American Honey Producers Association, have lobbied state and federal legislators and officials to oppose the release of a flea beetle, *Bikasha collaris*, which, according to USDA research, shows promise as a natural weapon against tallow trees.

"The introduction of this beetle for control of Chinese tallow would result in the loss of a major forage source for honey bees and other pollinator species . . . and lead to very serious economic impacts for beekeepers and others on a national scale," warned the Louisiana Beekeepers Association.

Perch, tamarisk, tallow: with so many species introduced to far-flung spots around the world, plenty have managed to find a useful role—at least from somebody's perspective.

Donkey Work

For millions of years, mammoths, ground sloths, and other enormous mammals known as "megafauna" stalked North America's forests and prairies. Exactly why nearly all of them vanished at the tail end of the last ice age remains a mystery, but scientists generally chalk their extinction up to some combination of rapid climate change and predation by humans. Regardless of what happened, when they disappeared in a relative blink of an eye these titans left holes in the ecological fabric, many of which would never be filled.

Osage orange trees (*Maclura pomifera*), large, thorny mulberry relatives, produce softball-sized green fruits full of sticky, bitter latex. Most animals today consider Osage orange fruits unpalatable, and to humans they're little more than a curiosity, known variously as "monkey balls" or "hedge apples." Eleven millennia ago, however, they would have been eaten whole by large mammals like woolly mammoths (*Mammuthus primigenius*), which dispersed the tree's seeds over long distances in their droppings. Without this service, the Osage orange can only disperse when its fruits float in streams or roll downhill, and since megafauna went extinct the wild range of this once-widespread species has contracted to a small area spanning northern Texas, southern Oklahoma, and western Arkansas. (Aging specimens also grow on agricultural land throughout the eastern United States, relics of a time when farmers planted them as living barbed wire.)

Dispersing seeds is only one of the so-called ecosystem services megafauna once performed. They also may have kept smaller herbivores in check, maintained tree-free grasslands, and even influenced fire regimes and soil microbes. And in the 2010s, ecologist Erick Lundgren discovered another long-vacant ecological role—although this time, it had been filled by an introduced species.

The horse family, *Equidae*, which also includes donkeys and zebras, evolved in North America roughly fifty-four million years ago. Ancestral horses dispersed into Asia, Europe, and Africa via the Bering Land Bridge before becoming extinct in

North America at the same time that ground sloths and mammoths disappeared. Roughly 10,500 years later, European settlers brought domesticated horses (*Equus equus*) and donkeys (*Equus africanus asinus*) back to North America for farmwork and transportation. Some escaped captivity or were released on purpose, and over time they formed self-sustaining feral populations in the American West, which generations of land managers have worked fruitlessly to control.

Lundgren was camping in Arizona when he first started noticing what he cheekily calls "ass holes," the wells dug by wild donkeys, known locally as burros, to access groundwater in arid environments. He was surprised to see that it wasn't only the asses who made use of these holes, which can be up to six feet deep, but also mountain lions (*Puma concolor*), mule deer (*Odocoileus hemionus*), migratory birds, and a variety of other native species. He returned to study the phenomenon in graduate school, first at Arizona State University and then at the University of Technology Sydney, where his advisor was Arian Wallach.

Lundgren's research showed feral donkeys, as well as horses, dramatically increase the amount of surface water available to other species. In some places he studied, their wells were the only surface water for miles around, making them literal lifesavers. They also provided water for plants, including flood-adapted species like cottonwoods, which were struggling in areas where humans had diverted streams for agriculture.

"In Africa, zebra and elephants dig water holes, and nobody's that surprised that these animals are so important to maintaining surface water, but nobody mentions it with the wild donkeys because they don't belong," Lundgren says in an interview from Arizona, where he is doing fieldwork. "The things that are really hard to spin as a negative impact, you just ignore."

Lundgren grew up among the woods of Upstate New York, developing an affinity for arid landscapes only as an adult. Among his treasures is a collection of Western curiosities: an early color postcard of two saddled burros, a cast of some surprisingly large coyote tracks, fossilized redwood leaves, a rusty pair of antique sheep shears. Each has its place in a glass-fronted cabinet with peeling paint, found fortuitously by the side of the road.

"I don't know why someone was throwing it out," he says.

With his wire-framed glasses and brown hair pulled back in a careless ponytail, Lundgren bears a distracting resemblance to David Foster Wallace. Like the late novelist he is also something of a non-conformist, and earnest to fault—characteristics that admittedly have not been great for his career. After earning his PhD in 2021, Lundgren took a postdoctoral position at Denmark's Aarhus University, then another three-year postdoc at the University of Alberta in Canada. All along, he has hunted fruitlessly for a tenure-track job.

"I can't be conclusive," he says. "But I've had several interviews at good schools. I have a very strong CV, good publications, and it's usually really positive—until I give a job talk." A standard part of the hiring process, the job talk is a chance for faculty hopefuls to dazzle the professors who may one day be their colleagues. Lundgren's don't tend to go well.

"I can see people visibly scowling and frowning, and then afterward they say really condescending things," he says. "The university hiring process is run by the department, so if you disagree, if you're critical of the ideas of people in the department, you're not going to get a job."

Lundgren can, in a way, sympathize with the scientists who find jarring any suggestion that non-native species might have an upside. As an undergraduate in Indiana, he volunteered with a local environmental center, where he spent much of his time eradicating invasive plants, finding like so many who participate in conservation efforts that pulling garlic mustard was a bit of a gateway drug.

"I realized, in myself, a kind of zealousness was emerging," he says. "And since then it's been kind of a long journey."

It was a conversation with his father that prompted Lundgren to question his own emotional response to invasive species.

"My dad asked, 'What will these species become?'" he remembers. "And that was the first time I thought, 'Oh wait. Yeah. Species have been moving around forever, and then they *become* the native species that we cherish.' I started looking at these plants with curiosity, which I think is the proper role of a scientist."

Lundgren began noticing aphids on garlic mustard, the insects seemingly ignorant of the prevailing narrative that it's basically inedible. In some areas, the plant didn't seem to be producing seeds—another surprise.

"Those types of dynamics I hadn't noticed before, because I was so hell-bent on killing it," he says. "We're told that nothing eats this, and it's basically invincible unless we humans go in there and remove it."

Despite a few high-profile heretics—like Lundgren—who still manage to place papers in top journals, such questioning remains largely anathema within universities and land management and conservation organizations, where Lundgren finds a "religious attitude" still prevails: "There's just something sacred about native species," he has heard people say, or, "Invasive species just don't belong."

Still, there's a growing recognition even in the mainstream that the simple invasion narrative may be flawed, and that strident, even dogmatic jargon offers an opening for critics. In a 2024 paper called "Taming the terminological tempest in invasion science," published in the journal *Biological Reviews*, an international group of several dozen researchers attempted to address controversy over the language of invasion biology. The authors, who included Daniel Simberloff and James Russell, reviewed common terms—"alien," "invader," "non-indigenous," "pest"—and proposed a "streamlined framework," translated into twenty-eight languages, based on the terms "non-native," "established non-native," and "invasive non-native," that take impact as well as origin into account.

"The lack of a clear terminology has been exploited in ongoing criticism from those who aim to undermine the value and fundamental goals of invasion science," the authors wrote.

Notably, while they acknowledged the difficulties of defining nativeness and the drawbacks of using emotionally laden terms, the proposed new framework preserves the central metaphor—and the term "invasive."

"There's a lot of middle-of-the-road ecologists that think that nativeness is overplayed, even if it matters sometimes, but these normative values are baked into the language so heavily," Lundgren says. "When you have metaphors like 'invasion,' it leads to an immediate image in our minds."

How Many Goodly Creatures

It remains to be seen whether humans can leave behind the metaphor of introduced species as invading armies and instead approach them with curiosity, on a case-by-case basis. But we might begin to ask ourselves, how long does it take for a species to naturalize? When can we consider it an ecological citizen in good standing? To echo the elder Mr. Lundgren, what *will* ultimately become of invasive species?

Over the past twenty years, some scientists and philosophers have begun to talk about "novel ecosystems." These are places in which humans have had a clear influence, introduced organisms can be counted as part of a place's biodiversity, and ecologists might study existing—rather than idealized—relationships among plants, animals, and people.

A widely cited 2006 paper by biologist Richard Hobbs, then at the University of Western Australia, and international colleagues, defined them as "ecosystems that are the result of deliberate or inadvertent human action, but do not depend on continued human intervention for their maintenance." They are abandoned pastureland in New England, European rivers modified by dams, and the tropical savannas of Brazil, which humans created with fire and maintain with grazing cattle. Arguably, they include the garbage middens that became tree islands in the Everglades, both the North American and the African Great Lakes, and even the seeming wilderness of the Australian Outback, where dingoes and rabbits now live alongside kangaroos.

Novel ecosystems are a controversial concept, one that mainstream invasion biologists reject out of hand as not only wrong but harmful. Hobbs et al. (2006) felt compelled to acknowledge this in their own paper, writing almost apologetically:

> This may seem to some to be a defeatist approach which recognizes that some ecosystems are more or less transformed irreversibly and that invasive species are likely to persist in some cases. Indeed, comments from reviewers of the draft manuscript indicated a lack of willingness to accept such ecosystems as a legitimate target for ecological thought or management action. For instance, one reviewer commented that the examples are ecological disasters,

> where biodiversity has been decimated and ecosystem functions are in tatters, and that 'it is hard to make lemonade out of these lemons'. . . . Our point is, however, that we are heading towards a situation where there are more lemons than lemonade, and we need to recognize this and determine what to do with the lemons.

Whether or not novel ecosystems are to be lauded or lamented, it's simply a neutral statement of fact that there are places all over the world—most places, by now—where species that can claim nativeness now live alongside others that have been brought by humans. And despite invasion biologists' feelings of despair—and enough clear-cut cases of harm done to justify trepidation—there is no universally applicable narrative, no sure forecast for how these relationships will shift and develop in the future.

For one thing, the fate of native species in novel ecosystems is not always to be consumed, superseded, and driven to extinction. Natives sometimes evolve faster than we think possible, adapting rapidly to the sudden appearance of an introduced predator, competitor, or food source—and even benefiting over the long term. In the Everglades, snail kites (*Rostrhamus sociabilis*) used to dine on the native Florida apple snail (*Pomacea paludosa*), but that mollusk got scarcer after the related South American island apple snail (*Pomacea maculata*) showed up in the early 2000s. Kites had already been through decades of habitat loss and drought, and scientists predicted doom for these endangered raptors, which initially struggled to eat the bigger snail.

In 2017, however, a team of researchers from the University of Florida found that kites had begun growing larger, with larger bills that enabled them to better extract the meat from the larger apple snails' shells. "Our findings . . . underscore that even long-lived vertebrates can respond quickly to invasive species," the authors wrote in the journal *Nature Ecology & Evolution*. By the 2020s, the kite population had rebounded, and some plucky birds were even expanding their territory as far away as North Carolina and Tennessee.

In 2020, the International Union for Conservation of Nature, the same body that compiles the Red List of Threatened Species, published a first edition of the Environmental Impact Classification for Alien Taxa, or EICAT, billed as a tool "for measuring the severity of environmental impacts caused by animals, fungi and plants living outside their natural range." Several years in the making, the framework offered categories for determining the impact of non-native species that ranged from "Minimal Concern . . . when it causes negligible levels of impacts, but no reduction in performance of individuals in the native biota," to "Massive," for those that cause "naturally irreversible community changes through local, sub-population or global extinction (or presumed extinction) of at least one native taxon." The particular "impact mechanisms" included not only competition and predation, but also hybridization, transmission of disease, and "bio-fouling," defined as "the accumulation of individuals of the alien taxon on the surface of a native taxon."

Notably, the framework has no room for the possibility of positive impacts, but some scientists have recently attempted to rectify what they believe is an oversight. In 2022 an international group of biologists proposed a framework they called "EICAT+" that added corresponding positive impacts all the way up to "massive," which might describe a scenario in which the introduction of a non-native species prevents a native from going extinct, perhaps by providing a new food source or by controlling a parasite or predator.

Under EICAT+, the island apple snail would likely get a reprieve for saving the Everglades snail kite. And these snails aren't just a godsend for the once-endangered raptor. Limpkins (*Aramus guarauna*), wading birds that feed mainly on aquatic invertebrates, were once listed as a species of special concern by the state of Florida, which hosted their only North American population. But with the arrival of island apple snails, the limpkin population boomed. Birders soon began reporting limpkin sightings in Louisiana, Texas, and even as far north as Ontario.

"Limpkins are everywhere all of the sudden. What is going on?" asked *Audubon* magazine, in 2024, citing trend data from the popular birdwatching app eBird that suggested the US limpkin population soared by more than 50 percent between 2012 and 2022. "Maybe it was the pressure of all those extra birds, but suddenly the cork popped, and the region wasn't big enough for them anymore."

Limpkins might have gotten a boost from the new snails, but they can survive in Canada only because the climate is growing warmer. In that, they join a who's-who of species moving north as winters become milder and summers warmer, a list that includes such diverse organisms as maple trees, lobsters, and butterflies. A 2024 review attributed 59 percent of all recently observed range shifts to climate change. This trend is going to make nativeness even harder to pin down, especially as it interacts with other anthropogenic changes, including ones we may never consider.

Humans, who have established permanent settlements in environments as diverse as the Sahara, the Amazon, and the Arctic, are probably the most widely distributed terrestrial species on Earth. But our competitors for the title are almost all species that have developed a close relationship with us, spreading to the remote outposts of the world alongside our ancestors. These include some of the most troublesome invasive species—rats, mice, cats—but also lower-profile commensals, like the barn swallow (*Hirundo rustica*).

Barn swallows originated in the Nile Valley, where they may have originally nested in caves. But when humans began building houses and barns, *H. rustica* discovered excellent new locations for their cup-shaped nests in the rafters and overhangs. As civilization radiated into the Middle East and throughout the Northern Hemisphere beginning roughly eleven thousand years ago, barn swallows followed, rapidly expanding their distribution and increasing their numbers.

"Our working hypothesis is that their distribution is a function of our own, because they basically live side by side with us," says Rebecca Safran, a professor of ecology and evolutionary biology at the University of Colorado Boulder, who has studied barn swallow genetics and breeding behavior for nearly three decades. Today, barn swallows eschew what we may think of as "natural" environments, instead settling everywhere from Canadian horse stables to Japanese railway stations: they are the ur-denizens of the novel ecosystem.

Safran picked barn swallows as her study subject largely because of their ubiquity: Wherever her career took her, she knew she would find them. She has researched how the six closely related subspecies of barn swallow differ from one another, how migratory behavior, plumage coloration, and the length of their tail streamers may be driven by sexual selection, and what happens when the populations hybridize. Genetic evidence suggests swallow subspecies have sometimes diverged and come back together over time. They have a "messy" evolutionary history, Safran says.

Today, *H. rustica*'s population is declining in many parts of the world, likely due to a combination of factors that includes overuse of pesticides, which kill the flying insects that swallows eat; modern builders' preference for steel and concrete over timber beams; and, inevitably, the same changing climate that is affecting everything else. Under these and other pressures, the subspecies are diverging once again and may be in the process of splitting entirely. Safran's youthful decision to study these birds, while originally based on practicalities, now has her well-placed to observe the origin of species in real time—and to philosophize about what that means.

"My perspective is of a very dynamic process," Safran says. "Population distributions change a lot over evolutionary history, and even from a modern time perspective, as boundaries go away and things move. Species are sort of a mirage at any given time point, and because evolution is constant, that mirage could disappear."

Darwin used the "tree of life" as a central metaphor for his theory of evolution by natural selection. Hierarchical, directional, and leading inevitably to greater complexity, it was an extremely useful concept, but one that was also bound to its time and to the values and patterns of thought common among Victorian scientists, influenced as they were by both Christian cosmology and the technology of the Industrial Revolution. In the 21st century, as modern genomics reveals new surprises—and as humans continue to toss wrenches into nature's works—the tree of life has turned out to be more of a bramble, tangling and doubling back on itself as species diverge and come together, spread and contract, prosper and vanish, and maybe reappear somewhere else.

From that perspective, the values of invasion biology become untenable, because the discipline presupposes a point in time when everything in nature was as it should be—but that moment could only ever be a screenshot from a movie that has no clear beginning or end.

"I remember once talking to a restoration ecologist," Safran says. "And my question for her was, 'What is our baseline for restoration? What is this biodiversity template that we aspire to restore to? Who gets to call that?' I understand the idea of one species getting incredibly successful, taking over, and becoming predominant at the expense of many others. I think humans really, really fit the bill there! But is that a problem, or is that just an outcome of evolutionary process?"

Unlike most of us, Safran is comfortable with uncertainty. All she can say for sure is that evolution is playing out, ebbing and flowing, with success begetting success—until for whatever reason that doesn't work anymore.

"At the end of the day, those are the patterns that we see," she says, "whether we like them or not."

It remains an open question, however, whether humans can bring ourselves to accept these changes and, in an era when technology may soon give us amazing new powers to reshape nature however we choose, whether we have to.

EPILOGUE

Halfway through the writing of this book, my family moved to the country.

During the COVID-19 pandemic, my husband and I had chafed against the walls of our small house in the city. We wanted more space, and we wanted it to be our space. We would grow a large garden of heirloom vegetables, plant some fruit trees, get some chickens, and let the kids play in the woods.

The ten acres where we settled are lovely, divided roughly in thirds between homestead, forest, and rolling meadow. The sunrise over the tree line is pink and golden, and spring mornings are a riot of birdsong—robins, wrens, rose-breasted grosbeaks, white-throated sparrows, indigo buntings, and an assortment of elusive warblers. After dark, there are fireflies and frog choruses and the occasional yip of a coyote. We planted the garden, and we did get the chickens, housing them in a sturdy, fortified coop to keep the carnivorous neighbors out. Old friends who come to visit from the city tell us it's idyllic.

What they don't know is that we've been thoroughly invaded.

At some point, decades ago, our place and the surrounding parcels were all part of a much larger farm, most likely one where people raised pastured livestock. We base this assessment mainly on a few remainders of wire fencing that poke out in the woods and on the smooth brome grass that dominates the meadow.

Yes, smooth brome (*Bromus inermis*) is an invasive species. It originated in Eurasia, but because it makes such good forage for sheep, horses, and cattle it was planted widely in the United States, continuing to thrive long after these pastures were abandoned. It's a perennial that spreads via underground rhizomes, and without chemicals, heavy machinery, or extended time under a heavy plastic tarp, it's impossible to kill.

We also have autumn olive, which seems to get bushier each year, and of course plenty of garlic mustard at the edges of the woods. In the spring, motherwort (*Leonurus cardiaca*), a perennial herb that also hails from Europe, is the first green shoot along the edge of the property line. There's Japanese barberry back among the oaks, and tangles of thorny multiflora rose. Asian bittersweet persists in climbing the green ash trees along the road, which aren't looking too healthy anyway—probably because Michigan's woods have been infiltrated by emerald ash borers (*Agrilus planipennis*), also from Asia. Dame's rocket (*Hesperis matronalis*) is at least pretty, with its clusters of pink and purple flowers, although its abundance is enough to make one uneasy. And while they're also brightly colored, we could definitely do without the Japanese beetles (*Popillia japonica*) that love to eat our garden plants.

Our neighbors to the east sit on twenty wooded acres, and in part because the woods we share are virtually impenetrable in the summer, we didn't meet them until a year after we had moved in. We were transplanting ramps (*Allium tricoccum*, or wild leek) under an oak tree, and Jim was motoring around on a four-wheeler, ripping up Japanese barberry. A scourge, he remarked, and we agreed—the thorns are hazardous, and there's one growing near my beloved ramp patch that bothers me (although knowledge of its long, yellow tap root, the hard shoveling required to excavate it, and the likelihood that it will regrow anyway has so far stayed my hand).

It was an unseasonably warm April afternoon, hot even in the woods, where the trees had yet to leaf out. We stood in the duff, comparing notes: Too much autumn olive. Too many deer. The ticks, the mosquitos. It was a pleasant chat, but I realized as Jim headed back to his barberry, followed closely by his friendly Siamese cat, that all we had talked about was the plants and animals we disliked and how we wished we could get rid of them. But what would be left if we did?

Until roughly twelve thousand years ago, this part of Michigan lay beneath a mile of ice. As the glacier retreated north, it carved out a landscape of rolling hills and kettle lakes that was eventually colonized by hardwood forest. But even as the ice melted, humans were moving into Michigan, camping along its shores, and hunting caribou and mastodons. Later, they would set up agricultural communities, growing squash, corn, and beans and modifying the landscape with fire to suit their needs.

When new settlers arrived from Europe, they trapped the beavers that had once engineered wetlands, chopped down the old growth forest and forced the original human inhabitants into reservations. The new landscape of stumps and brush burned, and then it burned again. Farmers planted row crops and pastured animals. Cities grew up and sprawled out. Some plants and animals were killed off or retreated under the new regime, while others—those darn deer!—thrived. Now the winters are becoming milder, the summers warmer. The ticks that carry Lyme disease seem to be taking advantage of the longer breeding season, along with the opossums that love to eat them, and the coyotes that might eat *them*.

It's a dizzying, at times dark and tangled history, one I sometimes wish I didn't know so much about. It would be easier to simply enjoy the lovely meadow, the pretty wildflowers, the clean air, and the quiet country life. But to love nature in the 21st century is to come to terms with a fallen world.

We're trying to do right by this land and to appreciate all the species with which we share it, these plants, animals, and fungi that thrive best in places humans have wrecked and remade. That's admittedly difficult with deer ticks (*Ixodes scapularis*), but it's easier with others, like stinging nettle (*Urtica dioica*), an Old World plant that grows by the brush pile and which makes a tasty spring vegetable in soups and side dishes. Or broadleaf plantain (*Plantago major*), imported by my European ancestors as a medicinal herb, which takes the sting away if you neglect to wear long gloves when harvesting nettles.

In the war over invasive species, one constant has been the use of religious metaphor by each side against the other. Perhaps mainstream invasion biologists are guilty of a misguided fervor to restore an antediluvian world. Perhaps compassionate conservationists are deluded by the false idol of animal personhood. Scientists might be "playing God" if they used genetics to erase an invasive population. The humans who move invasive species around—a sort of original sin—may have traversed boundaries that God himself established.

But which species belongs where is not really a question of good against evil, redemption against sin. It may be tempting, given everything that humans have done to the world, to see our power as almost godlike, and to argue over how we should wield it. It's harder to be humble, to recognize that nature is larger than any of us and that our species is a part of it, both acting and acted upon, trying to survive like any other, on a planet where the only thing we can be certain of is change.

ACKNOWLEDGMENTS

I would like to thank Veronica Goldstein, who believed in this book before anyone else did and whose expertise and emotional support have been invaluable to a first-time author at every step along the way. Thanks also to my editor, Jeremy Lewis, and to everyone who helped mold this into a real, physical book.

I am grateful to the scientists, land managers, and nature-loving citizens who shared their expertise, perspectives, and stories with me, including but not limited to Bernd Blossey, Victoria Nuzzo, Katie Grzesiak, Alycia Stigall, Juan David Carrillo, Mark Davis, Banu Subramaniam, Jim Hurley, Susan Roth, Marlene Condon, William Hamersky, Matthew Chew, Daniel Simberloff, Stephen Enloe, Traci Ardren, Radu Guiaşu, Premek Hamr, Mario Vallejo-Marin, Annalily van den Broeke, James Russell, Ally Palmer, Dan Henry, Jessi Morgan, Emily Major, Peter Marra, William Lynn, Daniel Ramp, Arian Wallach, Peter Wolf, Neil Gemmell, Damian Dowling, Erick Lundgren, and Rebecca Safran. I am especially grateful to LeRoy Rodgers, Brendon Hession, and Melissa Smith, who were so generous with their time.

A special thank you to Matthew Bertrand and all the staff and volunteers at the Washtenaw Parks and Recreation Department, who steward what may just be the best local park system in the country.

Thank you to the Thursday morning Zou Zou's crew, Michelle, Jeni, Marie-Anne and Monique, for their companionship and accountability, to Veronica Dreher for listening, and to all the friends who have asked after the project for years (yes, it's finally done!) and encouraged me along the way. And thanks, most of all, to my family, including my parents, Paul and Susan Crawford, who provided space and time to work, and to Dmitry, Ellie, and Avery, for their patience and support.

BIBLIOGRAPHY

INTRODUCTION

Diagne, C., et al. "High and Rising Economic Costs of Biological Invasions Worldwide." *Nature* (2021).

Giunti, Giulia, et al. "Invasive Mosquito Vectors in Europe: From Bioecology to Surveillance and Management." *Acta Tropica* (2023).

Hickman, Caroline, et al. "Climate Anxiety in Children and Young People and Their Beliefs About Government Responses to Climate Change: A Global Survey." *The Lancet Planetary Health* (2021).

Holling, C.S., and G.K. Meffe. "Command and Control and the Pathology of Natural Resource Management." *Conservation Biology* (1996).

Mather, Victor. "If You See This Creepy Fish, Cut Its Head Off." *New York Times*, April 10, 2025.

Pawson, Chad. "Taking Direct Action Against Invasive Species Is Helping These Volunteers Calm Their Climate Anxiety." CBC News, May 9, 2022.

Pecl, Gretta T., et al. "Biodiversity Redistribution Under Climate Change: Impacts on Ecosystems and Human Well-Being." *Science* (2017).

Rathore, Harshita. "The Super-Villian of the Plant Kingdom: Japanese Knotweed and Its Never-Ending Saga." *Rutgers Journal of Law & Public Policy* (Fall 2020).

Rising, Gerry. "The Questionable Wisdom of Introducing Alien Species." *The Kingbird* 51, no. 2 (June 2001).

Stateside Staff. "The Comeback Story of Michigan's Native Trumpeter Swans." Michigan Public Radio, May 18, 2021.

Tennant, Roy. "On the (Oddly Satisfying) War Against Invasive Species." *Bay Nature*, May 25, 2023.

Upham, Lailani. "Led by Tribal Scientists, Montana's Trumpeter Swan Revival Is a Triumph." *Audubon*, Fall 2018.

Wisconsin Department of Natural Resources. "DNR: State's Oldest Trumpeter Swan a Symbol of Success." *Post-Crescent*, January 11, 2015.

CHAPTER 1

Blossey, Bernd, et al. "Developing Biological Control of *Alliaria petiolata* (M. Bieb.) Cavara and Grande (Garlic Mustard)." *Natural Areas Journal* (2001).

Blossey, Bernd, et al. "Residence Time Determines Invasiveness and Performance of Garlic Mustard (*Alliaria petiolata*) in North America." *Ecology Letters* (2021).

Bossdorf, Oliver, et al. "Palatability and Tolerance to Simulated Herbivory in Native and Introduced Populations of *Alliaria petiolata* (Brassicaceae)." *American Journal of Botany* (2004).

Cipollini, Don, and Bill Gruner. "Cyanide in the Chemical Arsenal of Garlic Mustard, *Alliaria petiolata*." *Journal of Chemical Ecology* (2007).

Garcia-Quismondo, Manuel, J., et al. "Evolutionary Response of a Native Butterfly to Concurrent Plant Invasions: Simulation of Population Dynamics." *Science of the Total Environment* (2017).

Grieve, Margaret. *A Modern Herbal*. 2 vols. New York: Harcourt, Brace & Company, 1931.

Nuzzo, Victoria A. "Experimental Control of Garlic Mustard (*Alliaria officinalis*) in Northern Illinois Using Fire, Herbicide and Cutting." Report, Illinois Department of Natural Resources, August 1990.

Rodgers, Vikki L., et al. "Ready or Not, Garlic Mustard Is Moving In: *Alliaria petiolata* as a Member of Eastern North American Forests." *BioScience* (2008).

Rodgers, Vikki L., et al. "Where Is Garlic Mustard? Understanding the Ecological Context for Invasions of *Alliaria petiolata*." *BioScience* (2022).

CHAPTER 2

Argyrou, Vassos. *The Logic of Environmentalism: Anthropology, Ecology, and Postcoloniality*. New York: Berghahn Books, 2005.

Carrillo, J.D., et al. "Disproportionate Extinction of South American Mammals Drove the Asymmetry of the Great American Biotic Interchange." *Proceedings of the National Academy of* Sciences, 2020.

Carrillo, J.D., et al. "Neotropical Mammal Diversity and the Great American Biotic Interchange: Spatial and Temporal Variation in South America's Fossil Record." *Frontiers in Genetics* (2015).

Carson, Rachel. *Silent Spring*. Boston: Houghton Mifflin, 1962.

Darwin, Charles. *The Descent of Man, and Selection in Relation to Sex*. Princeton, NJ: Princeton University Press, 1981.

Darwin, Charles. *The Life and Letters of Charles Darwin, Including an Autobiographical Chapter*. New York: D. Appleton and Company, 1887. (Online at https://darwin-online.org.uk.)

Darwin, Charles. *The Origin of Species: 150th Anniversary Edition*. New York: Signet, 2003.

Davis, M.A. "Invasion Biology 1958–2005: The Pursuit of Science and Conservation." *In Conceptual Ecology and Invasions Biology: Reciprocal Approaches to Nature*, edited by M.W Cadotte, S.M. Mc. Mahon, T. Fukami, 35–64. London: Springer, 2006.

Elton, Charles S. *The Ecology of Invasions by Animals and Plants*. With a new foreword by Daniel Simberloff. Chicago: University of Chicago Press, 2000.

Harris, John A. "Like a Keen North Wind: How Charles Elton Influenced Silent Spring." *Science of the Total Environment* (2012).

Hernández, R., et al. "The Miocene La Venta Biome (Colombia)." *Geodiversitas* (2023).

Hody, James W., and Roland Kays. "Mapping the Expansion of Coyotes (*Canis latrans*) Across North and Central America." *ZooKeys* (2018).

Hooker, J.D. "Note on the Replacement of Species in the Colonies and Elsewhere." *The Natural History Review: A Quarterly Journal of Biological Science* (1864).

Krebs, Charles J. "Of Lemmings and Snowshoe Hares: The Ecology of Northern Canada." *Proceedings of the Royal Society B: Biological Sciences*, 2011.

Lear, Linda J. *Rachel Carson: Witness for Nature*. New York: Henry Holt, 1997.

Nagai, Kaori. "Rattus-Homo-Machine: Rats as Seafarers in the Nineteenth Century." In *Maritime Animals: Ships, Species, Stories*, edited by Kaori Nagai. University Park: Pennsylvania State University Press, 2023.

Southwood, Richard, and J.R. Clarke. "Charles Sutherland Elton. 29 March 1900–1 May 1991." *Biographical Memoirs of Fellows of the Royal Society*, 1999.

Stigall, A.L. "The Invasion Hierarchy: Ecological and Evolutionary Consequences of Invasions in the Fossil Record." *Annual Review of Ecology and Evolutionary Systematics* (2019).

Stigall, A.L., et al. "A Multidisciplinary Perspective on the Great Ordovician Biodiversification Event and the Development of the Paleozoic World." *Palaeogeography, Palaeoclimatology, Palaeoecology* (2020).

Webb, S.D. "The Great American Biotic Interchange: Patterns and Processes." *Annals of the Missouri Botanical Garden*, 2006.

Weissbroda, Lior, et al. "Origins of House Mice in Ecological Niches Created by Settled Hunter-Gatherers in the Levant 15,000 Years Ago." *Proceedings of the National Academy of Sciences*, 2017.

Winick, Stephen. "A Possum Crisp and Brown: The Opossum and American Foodways." *Folklife Today*, American Folklife Center & Veterans History Project, Library of Congress, August 15, 2019.

Wilson, Gregory P., et al. "A Large Carnivorous Mammal from the Late Cretaceous and the North American Origin of Marsupials." *Nature Communications* (2016).

CHAPTER 3

Achenbach, Joel. "Nature's Hostile Takeover." *Washington Post*, July 30, 2000.

Burdick, Alan. "Attack of the Aliens: Florida Tangles with Invasive Species." *New York Times*, June 6, 1995.

Cockburn, Andrew. "Weed-Whackers." *Harper's Magazine*, September 2015.

Dray, F. Allen, et al. "Invasion History of *Melaleuca quinquenervia* (Cav.) S.T. Blake in Florida." *Castanea* (2006).

Fairchild, David. "Reminiscences of Early Plant Introduction in South Florida." *Proceedings of the Florida State Horticultural Society*, 1938.

Fairchild, David. *The World Was My Garden: Travels of a Plant Explorer*. New York: Charles Scribner's Sons, 1938.

Fallows, James. "Immigration: How It's Affecting Us." *Atlantic*, November 1983.

Fanning, Timothy. "Killer Bees and the Sci-Fi Horror That Never Came to Pass in South Texas." *San Antonio Express-News*, June 15, 2023.

Forrest, George. "The Perils of Plant-Collecting." *The Gardeners' Chronicle: A Weekly Illustrated Journal of Horticulture and Allied* Subjects, May 21, 1910.

Geldmann, Jonas, and Juan P. González-Varo. "Conserving Honey Bees Does Not Help Wildlife." *Science* (2018).

Hyland, Howard L. "History of U.S. Plant Introduction." *Environmental Review* (1977).

Jones, Henry C. "How Plant Hunters Risk Lives to Find New Foods for Your Table." *Popular Science Monthly* (November 1922).

MacInnis, G.E., et al. "Decline in Wild Bee Species Richness Associated with Honey Bee (Apis mellifera L.) Abundance in an Urban Ecosystem." *PeerJ* (2023).

President's Committee of Advisors on Science and Technology (PCAST). *Teaming with Life: Investing in Science to Understand and Use America's Living Capital*. Washington, DC: US Government, 1998.

Subramaniam, Banu. *Ghost Stories for Darwin: The Science of Variation and the Politics of Diversity*. Urbana: University of Illinois Press, 2014.

United States Department of Agriculture, Bureau of Plant Industry. "Seeds and Plants Imported During the Period from December, 1903, to December, 1905." March 15, 1907.

United States Senate, Committee on the Judiciary. *Immigration Reform and Control: Report of the Committee on the Judiciary, United*

States Senate, on S. 2222, with Additional and Minority Views. 97th Cong., 2nd sess. Washington, DC: US Government Printing Office, 1982.

White, Theodore H. *America in Search of Itself: The Making of the President, 1956–1980*. New York: Harper & Row, 1982.

CHAPTER 4

Chew, Matthew. "A Theme Park Grows Beneath the Ground." *Boston Globe*, February 6, 2000.

Condon, Marlene A. "Blue Ridge Naturalist: Ecologists Recognizing Value of Alien Plants." *Crozet Gazette*, February 11, 2019.

Condon, Marlene A. "'Invasive' Plants Invaluable to Degraded Environment." *Crozet Gazette*, June 4, 2015.

Condon, Marlene A. "Rototilling/Gardening Efforts AND Mud Icing." *Crozet Gazette*, March 10, 2009.

Davis, Mark A. "Biotic Globalization: Does Competition from Introduced Species Threaten Biodiversity?" *BioScience* (2003).

Davis, Mark A. *Invasion Biology*. Oxford: Oxford University Press, 2009.

Davis, Mark A., and Kenneth Thompson. "Eight Ways to Be a Colonizer; Two Ways to Be an Invader: A Proposed Nomenclature Scheme for Invasion Ecology." *Bulletin of the Ecological Society of America* (2000).

Davis, Mark A., and Kenneth Thompson. "Newcomers 'Invade' the Field of Invasion Ecology: Question the Field's Future." *Bulletin of the Ecological Society of America* (2002).

Davis, Mark A., and Matthew Chew. "The Denialists Are Coming! Well, Not Exactly: A Response to Russell and Blackburn." *Trends in Ecology & Evolution* (2017).

Davis, Mark A., et al. "Don't Judge Species on Their Origins." *Nature* (2011).

Gross, Alan. *The Rhetoric of Science*. Cambridge, MA: Harvard University Press, 1990.

Guiaşu, R.C., and C.W. Tindale. "Logical Fallacies and Invasion Biology." *Biology & Philosophy* (2018).

IPBES. *Invasive Alien Species: Summary for Policymakers*. Bonn: IPBES Secretariat, 2023.

Laurance, William F. "Theory Meets Reality: How Habitat Fragmentation Research Has Transcended Island Biogeographic Theory." *Biological Conservation* (2008).

Lewin, Roger. "Santa Rosalia Was a Goat." *Science* (1983).

O'Connell, Maureen. "Freedom of Speech Shines in Arizona Cave." *High Country News*, July 3, 2000.

Ricciardi, Anthony, and Rachael Ryan. "The Exponential Growth of Invasive Species Denialism." *Biological Invasions* (2018).

Roth, Susan A., William Hamersky, and Manuel T. Lerdau. "To the Editor: The Blue Ridge Naturalist—Not." *Crozet Gazette*, March 12, 2019.

Russell, J.C., and T.M. Blackburn. "The Rise of Invasive Species Denialism." *Trends in Ecology & Evolution* (2017).

Sagoff, Mark. "Do Non-Native Species Threaten the Natural Environment?" *Journal of Agricultural and Environmental Ethics* (2005).

Sagoff, Mark. "Invasive Species Denialism: A Reply to Ricciardi and Ryan." *Biological Invasions* (2018).

Simberloff, Daniel. *Invasive Species: What Everyone Needs to Know*. Oxford: Oxford University Press, 2013.

Simberloff, Daniel, and E.O. Wilson. "Experimental Zoogeography of Islands: The Colonization of Empty Islands." *Ecology* (1969).

Simberloff, Daniel, et al. "*Non-Natives: 141 Scientists Object*." *Nature* (2011).

Thompson, Ken. *Where Do Camels Belong?: The Story and Science of Invasive Species*. London: Profile Books, 2014.

CHAPTER 5

Arden, Traci. "Human Influence on Tree Island Formation in the Florida Everglades: Lessons from Prehistory for Restoration Science." Paper presented at the 2018 International AMOC Science Meeting, Coconut Grove, Florida, 2018.

Buckingham, Gary R. "Biological Control of Alligatorweed, Alternanthera Philoxeroides, the World's First Aquatic Weed Success Story." *Castanea* (1996).

Cağlar, Sinan, and Dürdane Kolankaya. "The Effect of Sub-Acute and Sub-Chronic Exposure of Rats to the Glyphosate-Based Herbicide Roundup." *Environmental Toxicology and Pharmacology* (2008).

Cohen, Patricia. "Weed Killer, Long Deemed Safe, Is Blamed for Cancer." *New York Times*, September 20, 2019.

Finkl, Charles W., and Christopher Makowski. "The Florida Everglades: An Overview of Alteration and Restoration." In *Coastal Wetlands: Alteration and Remediation*, Coastal Research Library, vol. 21, edited by Charles W. Finkl and Christopher Makowski. Cham, Switzerland: Springer International Publishing AG, 2017.

Flores, Alfredo. "TAMEing Melaleuca with IPM." *AgResearch Magazine* (USDA Agricultural Research Service), November 2004.

Harris, Amanda. *Fruits of Eden: David Fairchild and America's Plant Hunters*. Gainesville: University Press of Florida, 2015.

Hu, Lin, Mingcong Chen, Xiaoran Xue, Mingyi Zhao, and Qingnan He. "Effect of Glyphosate on Renal Function: A Study Integrating Epidemiological and Experimental Evidence." *Ecotoxicology and Environmental Safety* (2025).

Keane, R.M., and M.J. Crawley. "Exotic Plant Invasions and the Enemy Release Hypothesis." *Trends in Ecology & Evolution* (2002).

Pandher, Upkardeep S., Shufang Li, Julia H. Charbonneau, et al. "Pulmonary Inflammatory Response from Co-Exposure to LPS and Glyphosate." *Environmental Toxicology and Pharmacology* (2021).

Wheeler, G.S., E.C. Lake, E. Mattison, and G.F. Sutton. "Host Range, Biology, and Climate Suitability of *Callopistria exotica*, a Potential Biological Control Agent of Old World Climbing Fern (*Lygodium microphyllum*) in the USA." *Biological Control* (2024).

CHAPTER 6

Bonnald, Julie, et al. "Who Are the Elephants Living in the Hybridization Zone? How Genetics May Guide Conservation to Better Protect Endangered Elephants." *Global Ecology and Conservation* (2021).

Crandall, Keith A., and Sammy De Grave. "An Updated Classification of the Freshwater Crayfishes (*Decapoda: Astacidea*) of the World, with a Complete Species List." *Journal of Crustacean Biology* (2017).

Crocker, Denton W., and David W. Barr. *Handbook of the Crayfishes of Ontario*. Toronto: University of Toronto Press, 1968.

de Queiroz, K. "Ernst Mayr and the Modern Concept of Species." *Proceedings of the National Academy of Sciences*, 2005.

Dobson, Andrew, Kezia Barker, and Sarah L. Taylor, eds. *Biosecurity: The Socio-Politics of Invasive Species and Infectious Diseases*. London: Routledge, 2013.

Fišer, Cene, et al. "Cryptic Species as a Window into the Paradigm Shift of the Species Concept." *Molecular Ecology* (2018).

Girard, Charles. "A Revision of the North American *Astaci*, with Observations on Their Habits." *Proceedings of the Academy of Natural Sciences of Philadelphia*, 1852.

Guiaşu, Radu Cornel. *Non-Native Species and Their Role in the Environment: The Need for a Broader Perspective*. Leiden: Brill, 2016.

Guiaşu, Radu Cornel, and Mark Labib. "The Unreliable Concept of Native Range as Applied to the Distribution of the Rusty Crayfish (*Faxonius rusticus*) in North America." *Hydrobiologia* (2021).

Hamr, Premek, et al. "The Distribution and Conservation Status of the White Colour Morph of *Faxonius propinquus* (Northern Clearwater Crayfish) in Lake Simcoe, Ontario, Canada." *Freshwater Crayfish* (2019).

Lopez, Jose V., et al. "Evolutionary Origins of the Invasive Marbled Crayfish (*Procambarus virginalis*)." *Journal of Crustacean Biology* (2017).

Lundy, Thomas. "Attack of the Clones: The Mysterious Mutant Threatening to Invade Canada's Waterways." *Canadian Geographic*, May 17, 2024.

Matos, Rubens, et al. "Phylogenomic Insights into the African Elephant Lineage." *Molecular Biology and Evolution* (2020).

Mellon, DeForest, Jr. "Smelling, Feeling, Tasting and Touching: Behavioral and Neural Integration of Antennular Chemosensory and

Mechanosensory Inputs in the Crayfish." *Journal of Experimental Biology* (2012).

Reynolds, Julian, and Catherine Souty-Grosset, eds. *Management of Freshwater Biodiversity: Crayfish as Bioindicators.* Cambridge: Cambridge University Press, 2011.

Rohland, Nadin, et al. "Genomic DNA Sequences from Mastodon and Woolly Mammoth Reveal Deep Speciation of Forest and Savanna Elephants." *PLoS Biology* (2010).

Simberloff, Daniel, et al. "U.S. Action Lowers Barriers to Invasive Species." *Science* (2020).

Simon, Chris. "There's Not Much We Can Do: How a Very Rare White Morph Crayfish Is Surviving in Lake Simcoe." *Barrie Advance*, May 8, 2023.

Stringer, Chris, and Lucile Crété. "Mapping Interactions of H. neanderthalensis and Homo sapiens from the Fossil and Genetic Records." *PaleoAnthropology* (2022).

Svardal, Hannes, et al. "Ancestral Hybridization Facilitated Species Diversification in the Lake Malawi Cichlid Fish Adaptive Radiation." *Molecular Biology and Evolution* (2020).

Thomas, John Rhidian, et al. "Terrestrial Emigration Behaviour of Two Invasive Crayfish Species." *Behavioural Processes* (2019).

Todesco, Marco, et al. "Hybridization and Extinction." *Evolutionary Applications* (2016).

Vallejo-Marin, Mario. "*Mimulus peregrinus* (*Phrymaceae*): A New British Allopolyploid Species." *PhytoKeys* (2012).

CHAPTER 7

Ascione, Frank. *Children & Animals.* West Lafayette, IN: Purdue University Press, 2005.

Fifield, Anna. "In New Zealand, a Soul-Searching Question: Are We as Open Minded as We Think We Are?" *Washington Post,* March 20, 2019.

Frankham, James. "Predator-Free Aotearoa: A Guide." *New Zealand Geographic,* July 2016.

Gallup. "Wellcome Global Monitor 2018: How Does the World Feel About Science and Health?" Commissioned by Wellcome. London: Wellcome Trust.

Gerolemou, R.V., et al. "Social Capital in the Context of Volunteer Conservation Initiatives." *Conservation Science and Practice*, 2022.

Gross, Rachel E. "New Zealand's War on 30 Million Possums." *Atlantic*, March 1, 2013.

Holm, Nicholas. "Consider the Possum: Foes, Anti-Animals, and Colonists in Aotearoa New Zealand." *Animal Studies Journal* (2015).

Howald, G., et al. "Invasive Rodent Eradication on Islands." *Conservation Biology* (2007).

Ihimaera, Witi. *The Amazing Adventures of Razza the Rat*. Auckland: Reed, 2006.

Kolbert, Elizabeth. "The Big Kill." *New Yorker*, December 15, 2014.

Linklater, Wayne, et al. "Predator Free 2050: A Flawed Conservation Policy Displaces Responsibility." *Conservation Letters* (2018).

Major, E. *Possums Are as Kiwi as Fish and Chips: Possum Advocacy and the Potential for Compassionate Conservation in Aotearoa New Zealand*. PhD diss., University of Canterbury, 2023.

Mulder, C.P.H., et al., eds. *Seabird Islands: Ecology, Invasion, and Restoration*. New York: Oxford University Press, 2011.

Owens, Brian. "Behind New Zealand's Wild Plan to Purge All Pests." *Nature* (2017).

Palmer, Alexandra, and Laura McLauchlan. "Landing Among the Stars: Risks and Benefits of Predator Free 2050 and Other Ambitious Conservation Targets." *Biological Conservation* (2023).

Potts, Annie. "Kiwis Against Possums: A Critical Analysis of Anti-Possum Rhetoric in Aotearoa New Zealand." *Society and Animals* (2009).

Russell, J.C., and N.D. Holmes. "Tropical Island Conservation: Rat Eradication for Species Recovery." *Biological Conservation* (2015).

Russell, J.C., et al. "Intercepting the First Rat Ashore." *Nature* (2005).

Russell, J.C., et al. "Predator-Free New Zealand: Conservation Country." *BioScience* (2015).

Sullivan, Helen. "Reality TV Contestant Apologises for Killing and Eating Protected New Zealand Bird." *Guardian*, July 24, 2024.

Sweetapple, Peter, et al. "Diet and Impacts of Brushtail Possum Populations Across an Invasion Front in South Westland, New Zealand." *New Zealand Journal of Ecology* (2004).

Trevino, Julissa. "South Georgia Island Is Officially Free of Its Bird-Killing Rodents." *Smithsonian Magazine*, May 10, 2018.

Young, A.M., T. Linné, and A. Potts. "Framing Possums: War, Sport and Patriotism in Depictions of Brushtail Possums in New Zealand Print Media." *Animal Studies Journal* (2015).

CHAPTER 8

Alley Cat Allies. "Alley Cat Allies Responds to Nature Study's Claims on Cats and Birds." Last modified January 30, 2013. https://www.alleycat.org/alley-cat-allies-responds-to-nature-studys-claims-on-cats-and-birds.

Alley Cat Allies. "Breaking Down the Bogus Smithsonian Catbird Study." Last modified July 2011. https://www.alleycat.org/resources/breaking-down-the-bogus-smithsonian-catbird-study.

Anker, Peder. *Imperial Ecology: Environmental Order in the British Empire, 1895–1945*. Cambridge, MA: Harvard University Press, 2001.

Balogh, A.L., et al. "Population Demography of Gray Catbirds in the Suburban Matrix." *Journal of Ornithology* (2011).

Bekoff, Marc, ed. *Ignoring Nature No More: The Case for Compassionate Conservation.* Chicago: University of Chicago Press, 2013.

Carey, John. "Cat Fight." *Anthropocene Magazine*, March 9, 2012.

Chew, Matthew. "Ecologists, Environmentalists, Experts, and the Invasion of the 'Second Greatest Threat.'" *International Review of Environmental History* (2015).

Coates, Peter. *Squirrel Nation: Reds, Greys and the Meaning of Home.* London: Reaktion Books, 2023.

Coman, Brian. *Tooth & Nail: The Story of the Rabbit in Australia.* Melbourne: Text Publishing Company, 2013.

Genovesi, Piero. "The Grey Squirrel in Italy: Risks of Expansion and Related Threats to the Survival of the Red Squirrel in Europe." *Proceedings of the Vertebrate Pest Conference*, 2000.

Genovesi, Piero. "Spread and Attempted Eradication of the Grey Squirrel (Sciurus Carolinensis) in Italy, and Consequences for the Red Squirrel (Sciurus Vulgaris) in Eurasia." *Biological Conservation* (2003).

Lynn, William S., Francisco J. Santiago-Avila, Joann Lindenmayer, John Hadidian, Arian Wallach, and Barbara J. King. "A Moral Panic over Cats." *Conservation Biology* (2019).

Lynn, William S., Francisco J. Santiago-Avila, and Kristen L. Stewart, eds. *Society & Animals: Special Issue on Outdoor Cats* (2022).

Loss, Scott R., and Peter P. Marra. "Merchants of Doubt in the Free-Ranging Cat Conflict." *Conservation Biology* (2018).

Loss, Scott R., et al. "The Impact of Free-Ranging Domestic Cats on Wildlife of the United States." *Nature Communications* (2013).

Loss, Scott R., et al. "Responding to Misinformation and Criticisms Regarding United States Cat Predation Estimates." *Biological Invasions* (2018).

Marra, Peter P., and Chris Santella. *Cat Wars: The Devastating Consequences of a Cuddly Killer*. Princeton, NJ: Princeton University Press, 2016.

McCure, Lily. "New 'Welfare Positive' Toxic Paste PAPPutty on Market to Target Fox, Wild Dog Populations." ABC News, May 22, 2024.

McNeely, Jeffrey A., ed. *The Great Reshuffling: Human Dimensions of Invasive Alien Species*. Gland, Switzerland: IUCN, 2001.

Nilson, S.M., et al. "Genetics of Randomly Bred Cats Support the Cradle of Cat Domestication Being in the Near East." *Heredity* (2022).

Peterson, Britt. "The Case Against Cats." *Atlantic*, December 2016.

Ratcliffe, F.N. "The Ecological Consequences of Myxomatosis in Australia." *Revue d'Écologie* (1956).

Rayner, Matt J., et al. "Spatial Heterogeneity of Mesopredator Release Within an Oceanic Island System." *Proceedings of the National Academy of Sciences*, 2007.

Ricciardi, Anthony, and Rachael Ryan. "Invasive Species Denialism Revisited: Response to Sagoff." *Biological Invasions* (2018).

Wallach, Arian, and Erick Lundgren. "Review of Evidence That Foxes and Cats Cause Extinctions of Australia's Endemic Mammals." *BioScience* (2025).

Wallach, Arian, et al. "Recognizing Animal Personhood in Compassionate Conservation." *Conservation Biology* (2020).

Wauters, L.A., et al. "Grey Squirrel Sciurus carolinensis Management in Italy - Squirrel Distribution in a Highly Fragmented Landscape." *Wildlife Biology* (1997).

Worster, Donald. *Nature's Economy: A History of Ecological Ideas*. Cambridge: Cambridge University Press, 1977.

Zukerman, Wendy. "Australia's Battle with the Bunny." *ABC Science*, April 8, 2009.

CHAPTER 9

Antão, L.H., et al. "Climate Change Reshuffles Northern Species Within Their Niches." *Nature Climate Change* (2022).

Brodie, Jedediah F., et al. "Shifting, Expanding, or Contracting? Range Movement Consequences for Biodiversity." *Trends in Ecology and Evolution* (2025).

Burt, Austin. "Site-Specific Selfish Genes as Tools for the Control and Genetic Engineering of Natural Populations." *Proceedings of the Royal Society B*, 2003.

Carrington, Damian. "UK Scientists to Launch Outdoor Geoengineering Experiments." *Guardian*, April 22, 2025.

Cattau, C.E., et al. "Effects of an Exotic Prey Species on a Native Specialist." *Biological Conservation* (2010).

Chew, Matthew. "The Monstering of Tamarisk: How Scientists Made a Plant into a Problem." *Journal of the History of Biology* (2009).

Gemmell, Neil J., et al. "The Trojan Female Technique: A Novel, Effective and Humane Approach for Pest Population Control." *Proceedings of the Royal Society B*, 2013.

Goldschmidt, Tijs. *Darwin's Dreampond: Drama on Lake Victoria*. Cambridge, MA: MIT Press, 1998.

Hobbs, Richard J., et al. "Novel Ecosystems: Theoretical and Management Aspects of the New Ecological World Order." *Global Ecology and Biogeography* (2006).

International Union for Conservation of Nature. *IUCN EICAT Categories and Criteria: The Environmental Impact Classification for Alien Taxa*, 1st ed. Gland, Switzerland: IUCN, 2020.

Kamadi, Geoffrey. "Secretive Traders Netting Chinese Delicacy Leave Nile Perch Under Threat." *Guardian*, May 27, 2019.

Lawlor, J.A., et al. "Mechanisms, Detection and Impacts of Species Redistributions Under Climate Change." *Nature Reviews Earth & Environment* (2024).

Lundgren, Erick, et al. "Equids Engineer Desert Water Availability." *Science* (2021).

Malewitz, Jim. "No Joke: Ann Arbor Is Removing Deer Ovaries. Lawmakers Aren't Laughing." *Bridge Michigan*, March 14, 2018.

Martin, William F., and Marek Mentel. "The Origin of Mitochondria." *Nature Education* (2010).

Michigan Department of Natural Resources. *Final Report on Sterilization of Game in Michigan: 2018 Public Act 390*. Lansing: Michigan Department of Natural Resources, May 26, 2020.

Novy, James E. "Screwworm Control and Eradication in the Southern United States of America." Food and Agriculture Organization of the United Nations, 1992.

Payne, Steve. "Louisiana Beekeepers Oppose Introduction of Beetle to Control Tallow Trees." *Louisiana Farm Bureau News*, January 15, 2018.

Pringle, Robert M. "Nile Perch." In *Encyclopedia of Biological Invasions*, edited by Daniel Simberloff and Marcel Rejmánek, 484–88. Berkeley: University of California Press, 2011.

Simberloff, Daniel. "Non-Native Invasive Species and Novel Ecosystems." *F1000Prime Reports* (2015).

Smith, Felisa A., et al. "Late Pleistocene Megafauna Extinction Leads to Missing Pieces of Ecological Space in a North American Mammal Community." *Proceedings of the National Academy of Sciences*, 2022.

Soto, Ismael, et al. "Taming the Terminological Tempest in Invasion Science." *Biological Reviews* (August 2024).

US Department of Agriculture, Agricultural Research Service. "First Biological Control Agent for Saltcedar Released." *ARS News & Events*, July 8, 1999.

Vimercati, Giovanni, et al. "The EICAT+ Framework Enables Classification of Positive Impacts of Alien Taxa on Native Biodiversity." *PLOS Biology* (2022).

Weidensaul, Scott. "Limpkins Are Everywhere All of the Sudden. What Is Going On?" *Audubon*, Fall 2024.

Wilkinson, Roger, and Gerard Fitzgerald. "Social Acceptability of the Trojan Female Technique for Biological Control of Pests." Melbourne: Department of Environment and Primary Industries, 2014.

Zhang, Sarah. "The 'Man-Eater' Screwworm Is Coming." *Atlantic*, May 27, 2025.

INDEX

For the benefit of digital users, indexed terms that span two pages (e.g., 52–53) may, on occasion, appear on only one of those pages.